Papaya

Biology, Cultivation, Production and Uses

Papaya

Biology, Cultivation, Production and Uses

Parmeshwar Lal Saran
Ishwar Singh Solanki
Ravish Choudhary

CRC Press
Taylor & Francis Group
Boca Raton London New York

CRC Press is an imprint of the
Taylor & Francis Group, an **informa** business

CRC Press
Taylor & Francis Group
6000 Broken Sound Parkway NW, Suite 300
Boca Raton, FL 33487-2742

Printed on acid-free paper
Version Date: 20150720

International Standard Book Number-13: 978-1-4987-3560-5 (Hardback)

Visit the Taylor & Francis Web site at
http://www.taylorandfrancis.com

and the CRC Press Web site at
http://www.crcpress.com

Contents

Foreword

Horticulture, including fruits, vegetables, medicinal and aromatic plants, mushrooms and floriculture, has emerged as an important sector for diversification of agriculture. Currently, horticulture has established its credibility in improving income through increased productivity, generating employment and enhancing exports, besides providing household nutritional security. The focused attention on enhancing investment in horticulture during the last decade has been rewarding in terms of increased production and productivity of horticultural crops with manifold export potential. India has emerged as the second largest producer of fruits in the world. Export of fruits and flowers has also increased many times over. The country already has a predominant share in the production of papaya.

India is the largest producer of papaya in the world, contributing about 37% with 106,000 hectares and 39.60 mt/hectare of productivity. Recent advances are finding increased applications in horticulture enzyme and pharmaceutical industries, and this trend should continue through crop improvement. Recent techniques to speed up genomics for development of viral disease–resistant and high-yielding varieties through genetic engineering are the need of the hour. Papaya has invariably improved the economic status of our farmers. The decision of the Indian Agricultural Research Institute Regional Station, Pusa, Bihar, to release a book on papaya is indeed very timely. It is expected to fulfill the long-felt need of all those engaged in papaya production and utilisation. To make this book useful to amateur farmers, all aspects related to this crop's cultivation, botany, biology, genetics, medicinal uses, protected cultivation, unfruitfulness, plant protection and physiological disorders have been covered in considerable detail.

With the vast available literature scattered over various sources, it is difficult for researchers and students to remain abreast of all the papaya industry–related developments. I compliment the commendable efforts of Dr. Parmeshwar Lal Saran, Dr. Ishwar Singh Solanki and Dr. Ravish Choudhary and congratulate them for their work on *Papaya: Biology, Cultivation, Production and Uses.* I am sure this publication will be immensely useful to researchers, teachers, students and other stakeholders in papaya cultivation and use.

R.S. Paroda
Trust for Advancement of Agricultural Sciences and
Former Secretary, DARE & DG, ICAR

Preface

Current climatic vulnerabilities place further stress upon new industrial approaches to a fruit sector facing alarmingly receding land and water resources, in addition to impeding population pressure leading to a food/fruit security crisis in the current global scenario. We cannot ensure sufficient fresh fruit supplies to feed teeming millions unless fruit technology is improved and modulated to keep pace with demand and supply. In the area of food and feed safety, this book covers the nutrients, anti-nutrients or toxicants, information of papaya's use as a food/feed and other relevant information. Papaya is one of the most nutritious and medicinally important fruit of the tropical region. Recently, its industrial value has also increased due to the enzyme 'papain', which has a number of industrial uses. Therefore, its cultivation has become highly profitable. Farmers can easily earn about Rs. 100,000 from a hectare of papaya plantation either by papain extraction or by marketing unripe and ripe fruits. Thus, our approaches toward fruit production, especially papaya farming, need to be redefined under subtropical conditions. Therefore, both academic and practical knowledge on papaya production is essential to formulate management practices for sustainable agricultural development. In providing fresh papaya produce and livelihood security to the masses of developing countries such as India, scientific papaya cultivation and efficient use of resources play key roles.

Thus, an attempt has been made to generate and compile the relevant information on recent advances in papaya production and to comprehend all aspects of managing this crop in *Papaya: Biology, Cultivation, Improvement and Uses*. This book comprises advanced information on agronomy, breeding, seed production technology, scientific crop management issues and protected cultivation. It also addresses compositional considerations for new varieties of papaya by identifying the key food and feed nutrients, anti-nutrients, toxicants and allergens. In addition to this, there is background material on the production, processing and uses of papaya and considerations to take into account when assessing new varieties of papaya. Constituents to be analysed related to food use and to feed use are suggested.

This book will be of immense use to students and faculty of the Indian Council of Agricultural Research (ICAR) and State Agricultural Universities (SAUs), and to researchers and academicians of other teaching, research- and extension-related organisations of regional, national and international repute. There may be deficiencies in this publication, which the authors/editors would like to improve upon in the future. Therefore, the authors welcome suggestions from readers to further improve the book.

Parmeshwar Lal Saran
Ishwar Singh Solanki
Ravish Choudhary

Acknowledgements

We are highly grateful to Dr. K.V. Prabhu, joint director (research), Indian Agricultural Research Institute, New Delhi, for successfully bringing about the publication of this book. His benevolent help, constant encouragement and great attitude were highly inspiring and motivating. We personally thank Dr. Mansa Ram and Dr. K. Singh, Indian Agricultural Research Institute, New Delhi, for encouraging us and extending their help in writing this book.

We acknowledge the contributions and assistance of Dr. P.R. Kumar, Dr. C.B. Singh, Rishi Raj and Dr. Md. Hashim, Indian Agricultural Research Institute, Regional Station, Pusa, Bihar, in the writing of this book. We also offer our sincere appreciation to Dr. P.K. Rai, Dr. Ranjeet Ray and Dr. S.K. Singh, Rajendra Central Agricultural University, Pusa, Bihar, for their helpful suggestions during the writing of this book. Special thanks to Dr. Ganga Devi, Department of Agricultural Economics, Anand Agricultural University, Anand, Gujarat, and Dr. Shashi Solanki, head of the Department of Chemistry, Chhaju Ram Memorial (CRM) Jat College, Hisar, Haryana, without whose encouragement, contributions and support this book would not have been written.

We would like to express our gratitude to all those who saw us through this process; to all those who provided support, talked things over, read, wrote, offered comments, allowed us to quote their remarks and assisted in the editing, proofreading and design. Thanks to CRC Press, Taylor & Francis Group, without whom this book would never have found its way to the Web and to the many people who cannot read this book. The real credit for this book ultimately goes to the many papaya researchers from the last 65 years who disclosed the necessary information. We are also very thankful to all the authors/publishers whose work was consulted and incorporated in the text.

Authors

Dr. Parmeshwar Lal Saran earned a master's in horticulture and was awarded a gold medal from the Rajasthan Agricultural University, Bikaner, Rajasthan, in 2002. He earned a PhD in fruit science from Chaudhary Charan Singh (CCS) Haryana Agricultural University, Hisar, Haryana, in 2005. He was a junior research officer (horticulture) at G. B. Pant University of Agriculture and Technology, Pant Nagar, starting in December 2006, and since March 2012 he has worked as a senior scientist (horticulture) at the Indian Agricultural Research Institute Regional Station, Pusa, Samastipur, Bihar. Dr. Saran received the Young Scientist Fellowship Award, 2014–2015, from the Department of Science and Technology, Science & Engineering Research Board, Government of India, New Delhi. He has authored more than 45 research publications in the field of horticulture.

Dr. Ishwar Singh Solanki earned a PhD in the genetics of plant breeding from Indian Agricultural Research Institute, New Delhi, in 1991. He has held the positions of assistant scientist, scientist and senior scientist (fruit breeding) at CCS Haryana Agricultural University, Hisar, Haryana, since September 1980. Since February 2009 he has worked as head of the Indian Agricultural Research Institute Regional Station, Pusa, Samastipur, Bihar. Dr. Solanki was bestowed the Best Teacher Award, CCS Haryana Agricultural University, Hisar, Haryana; Rashtriya Gaurav Award, IIFS, New Delhi; Fellow of ISG & PB, New Delhi; ISPRD, IIPR, Kanpur; and CHAI, New Delhi. He has released 14 varieties and registered 4 germplasm lines in agricultural crops, and has authored more than 100 research publications in field crops improvement.

Dr. Ravish Choudhary earned a master's in biotechnology in 2008 and a PhD in science from the University of Rajasthan, Jaipur, in 2012. Currently, he works as technical assistant at the Indian Agricultural Research Institute Regional Station, Pusa, Samastipur, Bihar. Dr. Choudhary has authored more than 25 research publications in relevant fields such as conservation and crop improvement.

1 Introduction and Uses

1.1 ORIGIN AND HISTORY

Papaya (*Carica papaya* L.), belonging to the family Caricaceae, is one of the most important fruits cultivated throughout the tropical and subtropical regions of the world. It is widely believed that papaya originated in the Caribbean coast of Central America, ranging from Argentina to Chile to southern Mexico through natural hybridisation between *Carica peltata* and another wild species (da Silva et al. 2007). Recently, another taxonomic revision was proposed and supported by molecular evidence that genetic distances were found between papaya and other related species (Kim et al. 2002). Some species that were formerly assigned to *Carica* family were classified in the genus *Vasconcella* (Badillo 2001). Accordingly, the classification of Caricaceae has been revised to comprise *Cylicomorpha*, *Carica*, *Jacaratia*, *Jarilla*, *Horovitzia* and *Vasconcella*, with *Carica papaya* being the only species within the genus *Carica* (Badillo 2001). The history of papaya appears to be first documented by Oviedo, the Director of Mines in Hispaniola (Antilles) from 1513 to 1525, where he described how Alphonso de Valverde took papaya seeds from the coasts of Panama to Darien, then to San Domingo and other islands of West Indies. The Spaniards gave it the name 'papaya' and took the plant to Philippines, from where it expanded to Malaya and finally India in 1598 (da Silva et al. 2007). By the time papaya trees were established in Uganda in 1874, their distribution had already spread through most tropical and subtropical countries.

When first encountered by Europeans, papaya was nicknamed 'tree melon'. Papaya is known by different names in the world, namely fafay and babaya (Arabic); thimbaw (Burmese); papayer and papaye (Creole); bisexual pawpaw, pawpaw tree, melon tree and papaya (English); papaya, lapaya and kapaya (Filipino); papailler, papaye and papayer (French); papaya and melonenbraum (German); gedang and papaya (Indonesian); kates (Javanese); doeumlahong (Khmer); Sino-Tibetan houng (Lao); papaali (Luganda); papaya, betek, ketalah and kepaya (Malaya); pepol (Sinhala); figuera del monte, frutabomba, papaya, papaita and lechosa (Spanish); mpapai (Swahili); ma kuaithet, malakor and loko (Thai); papayo (Tigrigna) and du du (Vietnamese). In Australia, red- and pink-fleshed cultivars are often known as 'papaya' to distinguish them from the yellow-fleshed fruits, known as 'pawpaw'. In India, it is locally known as pappaiya (Bengali), papeeta (Hindi), papaya (English) and pappali or pappayi (Tamil). Worldwide, India is the largest producer (5,160,390 MT), followed by Brazil (1517696 MT) (FAOSTAT 2012a, b) (Figure 1.1).

The market demand for tropical fruits has been growing steadily over the past two decades. Global production of tropical fruits (excluding bananas) reached 73.02 million (M) metric tonnes (t) in 2010. Gaining in popularity worldwide, papaya is now ranked third with 11.22 Mt, or 15.36% of the total tropical fruit production, after mango with 38.6 Mt (52.86%) and pineapple with 19.41 Mt (26.58%). Globally,

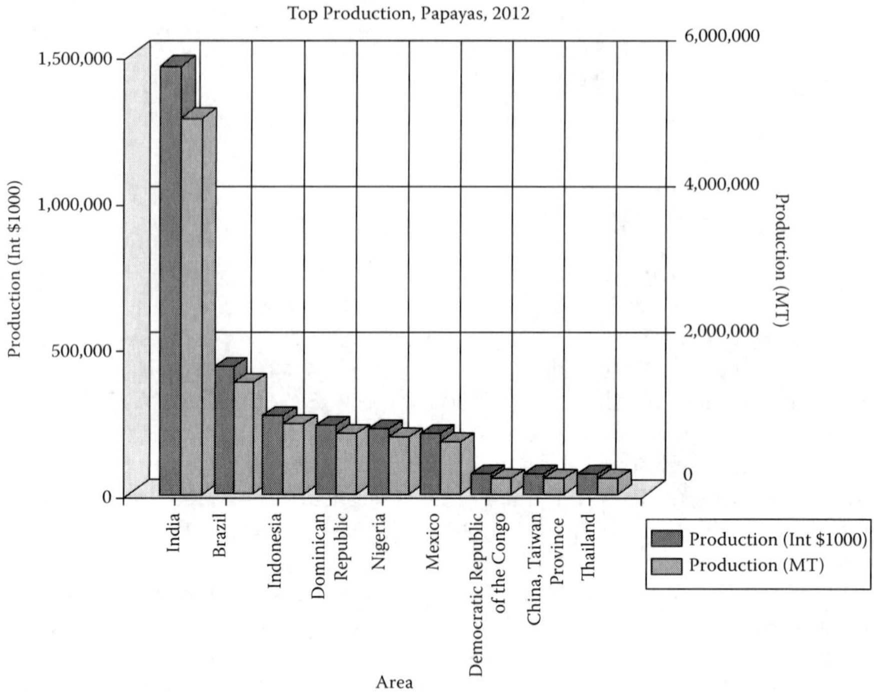

FIGURE 1.1 Production of papaya by region. (Adapted from FAOSTAT, 2012a, Papaya, Food and Agriculture Organization of the United Nations, faostat.fao.org.)

papaya production has grown significantly over the last few years, mainly as a result of increased production in India. Papayas are produced in about 60 countries, with the bulk of production occurring in the developing economies (Figure 1.2).

Global papaya production in 2010 was estimated at 11.22 Mt, growing at an annual rate of 4.35% between 2002 and 2010 (global production in 2010 was 7.26% higher than 2009 and 34.82% higher than 2002). Asia has been the leading papaya producing region, accounting for 52.55% of the global production between 2008 and 2010, followed by South America (23.09%), Africa (13.16%), Central America (9.56%), the Caribbean (1.38%), North America (0.14%) and Oceania (0.13%) (Figure 1.3) (FAOSTAT 2012b).

India is the leading papaya producer, with 38.61% share of the world production during 2008–2010, followed by Brazil (17.5%) and Indonesia (6.89%). While papaya production has remained relatively flat for most of the major producers, production in India has increased significantly within the past few years, and is chiefly responsible for the noticeable growth in global papaya production. From just over 2 Mt in 2005, papaya production in India has more than doubled to 4.7 Mt in 2010, representing an impressive annual growth rate of 14.94%. The biggest increase in global papaya production occurred between 2009 and 2010, as production in India increased by 20.50%. Such an impressive growth in production was due to a combination of increased acreage planted, improved genetics and better management (Evans and Ballen 2012). In India, it is cultivated in 73,000 ha with the production of 23.17 lakh tones (Singh et al. 2010). Major

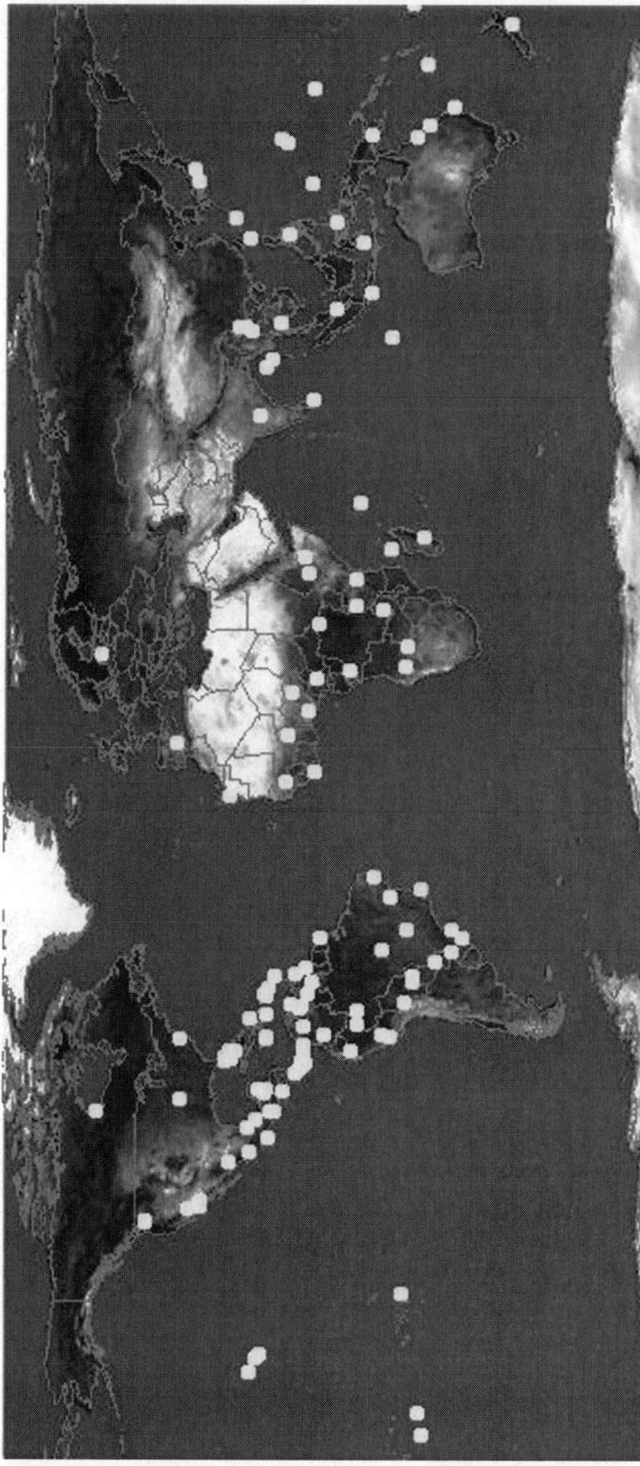

FIGURE 1.2 (See colour insert.) Major papaya growing countries. (From Encyclopaedia of Life (EOL), 2015, http://eol.org/pages/585682/maps.)

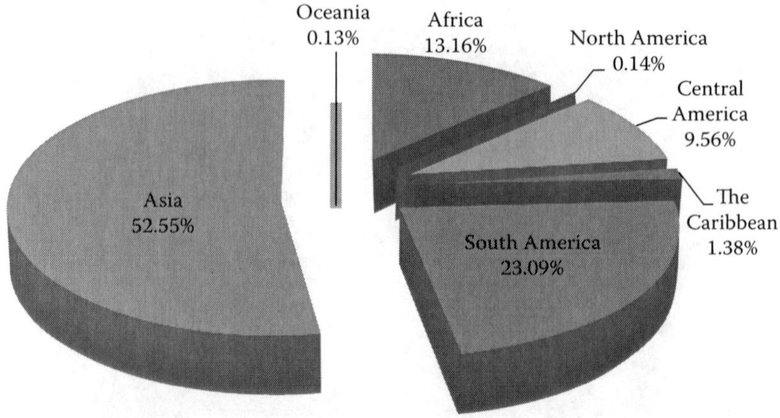

FIGURE 1.3 Papaya production by geographic area, 2008–2010. (Adapted from FAOSTAT, 2012b, Crop Production, http://faostat.fao.org/site/567/default.aspx#ancor; Fernando, J. A., M. Melo, and M. K. M. Soares, 2001, *Brazilian Archives of Biology and Technology,* 44:247–55.)

papaya growing states of India are Tamil Nadu, Gujarat, Karnataka, Maharashtra, Uttarakhand, Bihar, Jharkhand, West Bengal and Andhra Pradesh (Figure 1.4).

Papaya has gained more importance owing to its high palatability, fruitability throughout the year, early fruiting, highest productivity per unit area and multifarious uses like food, medicine and industrial input. Being highly remunerative and short duration fruit crop, it has a tremendous impact on economic and nutritional propitiations (Saran and Choudhary 2013). It needs plentiful rainfall or irrigation but must have good drainage. Papayas grow and produce well in a wide variety of soils. The tree often develops a strong taproot shortly after planting (Figure 1.5). The well-drained sandy loam soil with adequate organic matter is most important for papaya cultivation. Papaya fruit is consumed at both unripe and ripe stages. Unripe fruits are cooked and utilised as vegetables, processed products and as a source of papain (Mendoza 2007). Unripe papaya fruits are consumed both as a cooked vegetable and processed products (Morton 1987). Ripe papaya is consumed as a fresh fruit and also used for processing. At the unripe stage, the fruit is consumed as a cooked vegetable where papaya is widely grown (Mano et al. 2009), for example, in Thailand, unripe fruits are used as ingredients in papaya salad and cooked dishes (Sone et al. 1998) and in Puerto Rico, unripe fruits are canned in sugar syrup and sold either in local markets or exported (Morton 1987). The preserved unripe papaya fruit, which contains high sugar content, is used as an additive in ice cream. Green papaya fruit must be cooked (often boiled) prior to consumption to denature the papain in the latex (Odu et al. 2006). Ripe papaya fruits and papaya products, for their flavour and nutritional value, are consumed by humans (Saran 2010).

1.2 NUTRITIVE VALUE

The main constituent of *Carica papaya* fruit is water, like other fruits. The dry matter content increases during fruit development from unripe to ripe stages (Chavasit

FIGURE 1.4 Major papaya growing states of India.

et al. 2002). The agro-climatic conditions, cultivation practices, climate, seasons, site and cultivars, all these factors may influence the nutrient content of papaya (Hardisson et al. 2001; Wall 2006; Marelli de Souza et al. 2008; Charoensiri et al. 2009). Stages of maturity also affect the nutrient content of fruits like the vitamin C content of papaya increases with ripening (Bari et al. 2006). Consequently, when comparing the nutrient content of papaya fruits, it is important to compare fruits harvested and stored under similar conditions (Hernandez et al. 2006).

The major components of papaya dry matter are carbohydrates (USDA 2009). There are two main types of carbohydrates in papaya fruits, the cell wall polysaccharides and soluble sugars. During the early stage of fruit development glucose is the main sugar. The sucrose content increases during the ripening process and can reach levels up to 80% of total sugars (Paull 1993). Among major soluble sugars in ripe fruits (glucose, fructose and sucrose), sucrose is the most prevalent. During fruit ripening, the sucrose content was shown to increase from 13.9 ± 5.0 mg/g fresh weight in green fruit to 29.8 ± 4.0 mg/g fresh weight in ripe fruits (Gomez et al. 2002). The total dietary fiber content in ripe fruits varies from 11.9 to 21.5 g/100 g dry matter (Wills et al. 1986; Puwastien et al. 2000; Saxholt et al. 2008). The

FIGURE 1.5 (**See colour insert.**) Mature papaya tree cv., Pusa Dwarf.

crude protein content ranges from 3.74 to 8.26 g/100 g dry matter, and aspartic acid is the most abundant amino acid in ripe fruits, followed by glutamic acid. The chemists of Italy and Somalia collaboratively identified 18 amino acids in seeds as given in descending order, namely glutamic acid, arginine, proline and aspartic acid in the endosperm, and proline, tyrosine, lysine, aspartic acid and glutamic acid in the sarcotesta. The major organic acids found in ripe papaya are citric acid (335 mg/100 g FW), followed by L-malic acid (209 mg/100 g FW), qiunic acid (52 mg/100 g FW), succinic acid (52 mg/100 g FW), tartaric acid (13 mg/100 g FW), oxalic acid (10 mg/100 g FW) and fumaric acid (1.1 mg/100 g FW) (Hernandez et al. 2009).

Papaya is regarded as an excellent source of vitamin C (ascorbic acid), a good source of carotene and riboflavin, and a fair source of iron, calcium, thiamin, niacin, pantothenic acid, vitamin B-6 and vitamin K (Bari et al. 2006; Adetuyi et al. 2008; Saxholt et al. 2008). Carotenoid content (13.80 mg/100 g dry pulp) of papaya is low compared to mango (50–260 mg/100 g dry pulp), carrot and tomato (Saran 2010). The major carotenoid is cryptoxanthin. Carotenoids are responsible for the flesh colour of papaya fruit mesocarp. Red-fleshed papaya fruits contain five carotenoids, namely beta-carotene, beta-cryptoxanthin, beta-carotene-5-6-epoxide, lycopene and zeta-carotene. Yellow-fleshed papaya contains only three carotenoids, namely beta-carotene, beta-cryptoxanthin and zeta-carotene (Chandrika et al. 2003). Papaya is a source of vitamin C with amounts varying between the maturation stages (Bari et al., 2006; Hernandez et al. 2006). The total lipid content in ripe papaya fruit varies between 0.92 and 2.2 g/100 g dry matter. Papaya contains a low level of fatty acids. Palmitic acid and linolenic acid are two major fatty acids in papaya. Fatty acid composition changes during fruit ripening and no significant differences are observed in lipid composition with maturity of papaya fruits. The lowest and the highest levels

TABLE 1.1
Food Value per 100 g of Edible Portion of Papaya

Ingredient	Fruit	Leaves
Calories	23.10–25.80	–
Moisture	85.90–92.60 g	83.30%
Protein	0.081–0.34 g	5.60%
Fat	0.05–0.96 g	0.40%
Carbohydrates	6.17–6.75 g	8.30%
Crude fibre	0.50–1.30 g	1.00%
Ash	0.31–0.66 g	1.40%
Calcium	12.90–40.80 mg	0.406% (CO)
Phosphorus	5.30–22.00 mg	–
Iron	0.25–0.78 mg	0.0064%
Carotene	0.005–0.676 mg	28,900 I.U.
Thiamine	0.021–0.036 mg	–
Riboflavin	0.024–0.058 mg	–
Niacin	0.23–0.55 mg	–
Ascorbic acid	35.50–71.30 mg	38.60%
Tryptophan	4.00–5.00 mg	–
Methionine	1.00 mg	–
Lysine	15.00–16.00 mg	–
Magnesium	–	0.035%
Phosphoric acid	–	0.225%

of constituents in ripe papaya fruit and current mature leaves are given in Table 1.1. The edible portion of fruit contains macrominerals like, sodium, potassium, calcium, magnesium and phosphorus. The microminerals include iron, copper, zinc, manganese and selenium (USDA 2009).

1.3 USES

1.3.1 Fruit

Ripe papaya fruit is most commonly consumed like a melon. It can be peeled, seeds removed, cut into pieces and served as a fresh fruit. It can also be cut into wedges and then served with lime or lemon. Ripe papaya is also used in jam, jelly, marmalade and other products containing added sugar. Other processed products include puree or wine, nectar (Matsuura et al. 2004; Nwofia et al. 2012), juice, frozen slices or chunks, mixed beverages, papaya powder, baby food, concentrated and candied items (OECD 2005, 2008; OGTR 2008). Papaya puree is prepared from fully ripe peeled fruit with the seeds removed. Papaya flesh is pulped, passed through a sieve and thermally treated. Papaya puree is an important intermediate product in the manufacture of several products such as beverages, ice cream, jam and jelly (Brekke et al. 1972; Ahmed et al. 2002). Papaya nectar is prepared from papaya puree and consumed either alone or with other fruit juices such as passion fruit juice and pineapple juice (Brekke et al. 1972). It should be stored at or below 24°C to maintain acceptable quality (Brekke et al. 1976). Ripe papayas are most commonly eaten fresh, merely peeled, seeded, cut into wedges and served with a half or quarter of lime or lemon. Sometimes, a few seeds are left attached for those who enjoy their peppery flavour but not many should be eaten because the seed extract of papaya causes sterility in mammals also. The flesh is often cubed and served in fruit salad or fruit cup. Firm-ripe papaya may be seasoned and baked for consumption as a vegetable. Ripe flesh is commonly made into sauce for shortcake or ice cream sundaes, or is added to ice cream just before freezing, or is cooked in pie, pickled or preserved as marmalade or jam. Papaya and pineapple cubes, coated with sugar syrup, are quick frozen for later serving as dessert. Half-ripe fruits are sliced and crystallised as a sweetmeat for consumption. Unripe papaya is never eaten raw because of its latex content. Raw green papaya is frequently used in Thai and Vietnamese cooking. Even for being used in salads, it must first be peeled, seeded and boiled until tender, then chilled. Green papaya is frequently boiled and served as a vegetable. Cubed green papaya is cooked in mixed vegetable soup. Green papaya is commonly canned in sugar syrup in Puerto Rico for local consumption and export. Green papayas for canning in Queensland must be checked for nitrate levels. High nitrate content causes damage to ordinary cans, and all papayas with over 30 ppm nitrate must be packed in cans lacquered on the inside. Australian growers are hopeful that papaya can be bred for low nitrate uptake. A lye process for batch peeling of green papayas has proven feasibility in Puerto Rico. The fruits may be immersed in boiling 10% lye solution for 6 min, in a 15% solution for 4 min or a 20% solution for 3 min. They are then rapidly cooled by cold water bath and then sprayed with water to remove all softened tissue. Best proportions are 0.45 kg of fruit for every 3.8 L of solution.

Drying and freeze drying are used to reduce the moisture content of papaya chunks and slices. Powdered or dried papaya can be used as a flavouring agent, meat tenderiser or as an ingredient in soup mixes (Singfield 1998). Papaya seeds are a good source of 18 amino acids and edible oil. Seeds are sometimes also used to adulterate whole black pepper (Morton 1987).

1.3.2 Leaves, Pomace and Fruit Skin

Young leaves are cooked and eaten like spinach in the East Indies. Mature leaves are bitter and must be boiled with a change of water to eliminate much of the bitterness. Crushed leaves may be used to tenderise meat; however, stomach trouble, purgative effects and abortion may result from consumption of the dried papaya leaves (Morton 1987). Papaya leaves contain bitter alkaloids, carpaine and pseudocarpaine, which act on the heart and respiration like digitalis, and can be destroyed by heat. In addition, two previously undiscovered major piperideine alkaloids, dehydrocarpaine I and II, more important than carpaine, were reported. Papaya pomace, skins, leaves and other by-products of papaya processing may find use in animal feed applications (Fouzder et al. 1999; Alobo 2003; Babu et al. 2003; Reyes and Fermin 2003; Ulloa et al. 2004).

1.3.3 Flower and Stem

Sprays of male flowers are sold in Asian and Indonesian markets and in New Guinea after boiling with several changes of water to remove bitterness and then eating as a vegetable. In Indonesia, the flowers are sometimes candied. Young stems are cooked and served in Africa. Older stems are grated after peeling, the bitter juice squeezed out, and the mash mixed with sugar and salt (Saran and Choudhary 2013).

2 Botany and Improvement

2.1 PLANT MORPHOLOGY

The papaya plant is a semi-woody, latex-producing, usually single-stemmed, short-lived perennial herb. The relatively small genome of this species shows peculiarities in major gene groups involved in the cell size and lignification, carbohydrate economy, photoperiodic responses and secondary metabolites, which place papaya in an intermediate position between herbs and trees (Ming et al. 2008). Reproductive precocity, high photosynthetic rates of short-lived leaves, fast growth, high reproductive output, production of many seeds and low construction cost of hollow stems, petioles and fruits characterise this successful tropical pioneer (Ming et al. 2008). High phenotypic plasticity allows this plant to establish in recently disturbed sites, thriving during early stages of tropical succession and as member of diverse agro-ecosystems as well (Ewel 1986), that constitute important genetic reservoirs (Brown et al. 2012). Under appropriate conditions of water availability, light, oxygen, air temperature and humidity, papaya seeds undergo epigeal germination; emergence is typically completed in 2–3 weeks (Fisher 1980). The papaya plant develops very fast, taking 3–8 months from seed germination to flowering (juvenile phase) and 9–15 months for harvest (Paterson et al. 2008). The plant can live up to 20 years; however, due to excessive plant height and pathological constraints, the commercial life of a papaya orchard is normally 2–3 years. At any given time, adult papaya plants can sustain vegetative growth, flowering and dozens of fruits at different stages of development simultaneously (Ming and Moore 2014).

2.1.1 Stem

Papaya is a fast-growing arborescent herb with a short life; it has single, straight or sometimes, branched stem reaching 2–10 m in height. The stem is cylindrical, spongy fibrous, loose, hollow, grey or grey-brown in colour, 10–30 cm in diameter and toughened by large and protuberant scars caused by fallen leaves and flowers. Occasionally, vigorous vegetative growth may induce axillary bud break and branching at the lower portions of the plant, which rarely exceeds a few centimetres in length. Some branching may also occur if apical dominance is lost due to tip damage, and in tall plants, 'distance' may release the lower buds from the dominant effect of the apex (Morton 1987).

2.1.2 Leaf

Primary leaves of young seedlings are not lobed but become so after the appearance of the second leaf. Leaves are alternate, bundled at the apex between stem and branches, long petioles; widely evident, 40–60 cm diameter (Ming et al. 2008), smooth, moderately palm shape with thick middle irradiant veins, the base is deeply string shape with over imposed lobes; from 7 to 11 large lobed, each with a wide base or slightly constrained and sharp-pointed, and sharp apex. The bundle of leaves is dark green to yellowish green, bright, visibly marked by the off-white nerves embedded and reticulated veins; the underneath surface is pale green-yellow and opaque with visibly prominent vascular structures. Leaf blades are dorsiventral and subtended by 30–105 cm long, hollow petioles that grow nearly horizontal (25–100 cm length and 0.5–1.5 cm thick), endowed with a starch-rich endodermis, perhaps important for cavitation repair (Leal-Costa et al. 2010). The leaf epidermis and the palisade parenchyma are composed of a single cell layer, while the spongy mesophyll consists of four to six layers of tissue. Reflective grains and druses are abundant throughout the leaf (Fisher 1980). Papaya leaves are hypostomatic with anomocytic (no subsidiary cells) or anisocytic (asymmetric guard cells) stomata (Leal-Costa et al. 2010). Stomatal density of sunlit leaves is approximately 400/mm^2, which can adjust readily to environmental conditions of light, water and heat. Important biologically active compounds have been identified in papaya leaves, where they function in metabolism, defence, signalling and protection from excess light, among others (Konno et al. 2004; Ming and Moore 2014).

2.1.3 Root

Papaya is very susceptible to wind break due to its flat root system, especially in the monoculture. The papaya root is predominately a non-axial, fibrous system, composed of one or two 0.5–1.0 m long tap roots. Secondary roots emerge from the upper sections and branch profusely. These second-order feeding roots remain shallow during the entire life of the plant and show considerable gravitropic plasticity. Many adventitious, lower order categories of thick and fine roots are also observed in excavated specimens. Healthy roots are of whitish cream colour, and no laticifers have been observed in them (Carneiro and Cruz 2009). The root phenotypic plasticity is also high. The root size, number, distribution and orientation adjust readily across the soil profile, to various soil conditions and throughout the life of the plant, making papayas preferred components of complex agro-ecological models and hillside vegetation (Marler and Discekici 1997).

Leaves, stems and roots of young papaya plants exhibited rapid adaptive responses to wind. Wind stress of only one week was sufficient to elicit significant responses. Stem responses were the most rapid, but the root tip density responses exhibited the greatest magnitude among all wind conditions. The asymmetry index is translated to the windward side exhibiting 1.8–1.9 times more root tips than the leeward side. The form of a root system becomes increasingly variable with the age of tree as it responds to a number of stimuli and the bulk of the plant body increases (Marler 2011).

2.1.4 Flower

Plants are dioecious or hermaphroditic, with cultivars producing only female or bisexual (hermaphroditic) flowers. Papayas are sometimes said to be ‘trioecious’ meaning that separate plants bear either male, female or bisexual flowers. The female and bisexual flowers are waxy, ivory white and borne on short peduncles in leaf axils along the main stem. Flowers are solitary or small cymes of three individuals. Ovary position is superior. Prior to the opening, bisexual flowers are tubular and female flowers are pear-shaped. Since bisexual plants produce the most desirable fruit and are self-pollinating, they are preferred over female or male plants. A male papaya is distinguished by the smaller flowers borne on long stalks. Female flowers of papaya are pear shaped when unopened, and distinguished from bisexual flowers which are cylindrical. Bisexual flowered plants are self-pollinating, but flowers on female plants must be cross-pollinated by either bisexual or male plants flowers (Morton 1987).

2.1.5 Fruit

Individual fruits mature in 5–9 months, depending on cultivar and temperature. Plants begin bearing in 6–12 months. Fruits are large, oval to round berries, sometimes, called pepo-like berries since they resemble melons by having a central seed cavity. Fruit are borne axillary on the main stem, usually singly, but sometimes, in small clusters. The fruit is 15.0–50.0 cm long and 10.0–20.0 cm thick; weighing up to 9.0 kg (Morton 1987). The skin is waxy and thin but fairly tough. When the fruit is immature, it is rich in white latex and the skin is green and hard. As ripening progresses, papaya fruits develop light or deep yellow-orange coloured skin, while the thick wall of succulent flesh becomes aromatic, yellow orange or various shades of salmon or red. It is then juicy, sweetish and somewhat like a cantaloupe in flavour, but sometimes is quite musky (Morton 1987). Mature fruits contain numerous grey-black ovoid seeds attached lightly to the flesh by soft, white and fibrous tissue. These corrugated, peppery seeds of about 5.0 mm in length are each coated with a transparent and gelatinous aril (Jaime et al. 2007).

2.1.6 Seed

Well-pollinated fruits result in 400–600 seeds per fruit. The embryo is straight and ovoid, with flattened cotyledons. The seeds are coated by a mucilaginous mass derived from the pluristratified epidermis of the outer integument. The embryo is enclosed in a gelatinous layer (aril) known as sarcotesta at physiological maturity (Ming and Moore 2014). A compact mesotesta and outer and inner integuments can be observed underneath. The endosperm is composed of thin-walled cells with abundant oil bodies and aleuronic grains, lacking starch at maturity (Teixeira da Silva et al. 2007). Photosensitive seeds of wild papayas are dormant at maturity, and their germination may be triggered by changes in light quality (Paz and Vazquez-Yanes 1998).

Papaya seeds generally have been classified as the intermediate seed owing to intermediate seed storage capacities and storage for periods greater than 5 years is difficult (Ellis et al. 1991; Wood et al. 2000). Loss of viability has been reported at moisture contents below 8.0%–10.0% (Ellis et al. 1990), although viability has been reported when seed was desiccated to 5.0% moisture contents (Magill et al. 1994).

However, papaya seeds have been grouped as recalcitrant type by Chin et al. (1984) and Hofmann and Steiner (1989). Propagation of papaya by seed is difficult due to rapid seed deterioration after harvest. This is attributed to microbial degeneration of the sarcotesta which reduces viability. Salomao and Mundim (2000) have reported an orthodox behaviour. Due to dehydration up to 5.3%–6.9% moisture content (MC), followed by exposure to sub-zero temperatures and treatment with GA_3 were the most favourable combined treatments to enhance papaya seed germination. This type of seed nature has permitted germplasm conservation in conventional and cryogene banks. Wood et al. (2000) have reported that dormancy results from desiccation of papaya seeds and that desiccation-induced dormancy can be reversed by heat shock. Poor germination of papaya seeds after drying to 4.5%–11.5% MC is due to the induction of dormancy rather than a loss in viability. Desiccation-induced dormancy can be removed by heat-shocking rehydrated seeds for 4 h at 36°C before return to a 26°C germination regime. Both the endosperm and embryo of intermediate seeds of papaya display an O:S ratio within the range of values found for orthodox seed tissues.

2.2 FLORAL MORPHOLOGY

The inflorescence of a male plant is long, pendulous panicle bearing many flowers. Male flowers are small, about 2.0–3.0 cm long, and have a minute calyx of five small united sepals, and a long corolla and tube divided about one-third of the way from its mouth into five pointed lobes. The pale yellow corolla is often fragrant; at the mouth of the tube 10 epipetalous stamens are arranged in two rows of five, one row with longest filament lying opposite with the petal lobes, the other with shorter filaments lying opposite the lobes (Figure 2.1).

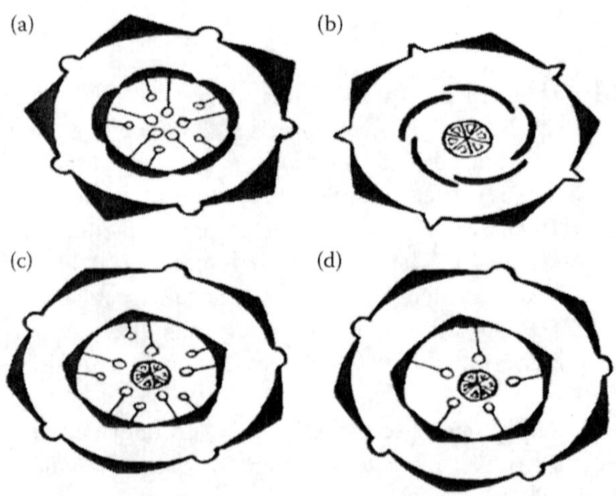

FIGURE 2.1 Floral diagrams: staminate (a), pistillate (b), normal hermaphrodite (c) and elongata hermaphrodite (d).

Female flowers are much larger than male flowers, around 4.0–5.0 cm long, and are more or less sessile on the main axis. They have a short calyx-tube with five short lobes, and five waxy, yellow petals united at the base but free for most of their length and with twisted, pointed tips. The large superior ovary is sessile, globular and green with five much branched sessile stigmas. It is composed of five united carpels with a single locule which contains many ovules on parietal placenta. They are produced in cymes 2.5–10.0 cm long, bearing one to six flowers. All the female flowers are purely pistillate without any visible evidence of stamen vestiges. This form of tree does not respond to the physiological influences that cause sex reversals in certain hermaphrodite and male trees.

The hermaphrodite tree has cymes which are 2.5–25.0 cm long, bearing one to six flowers. Usually, the terminal flowers of the primary and secondary peduncles are perfect. Flowers from tertiary and lower orders of peduncles may be perfect or staminate. Most hermaphrodite plants are subject to 'sex reversal' or modification. The change of sex occurs, from male to female, but never from female to male.

The hermaphrodite flower in case of 'elongata' has elongate ovary which develops into a more or less cylindrical, pyriform or oblong fruit depending on cultivar or strain. This is commonly referred to as the perfect flower. Its calyx is typical of the calyxes of all papaya flowers. The corolla is large, more or less fleshy, and sympetalous for one half to three fourth of its length. It has 10 versatile introse stamens. These are coalesced into a monadelphous ring found at throat of the corolla tube. Typically, the pistil is five-carpellate and uniloculated, with parietal placentation. In some flowers, however, the number of carpels may lie between one and ten. The fruit develops from a superior ovary.

A reduced type of hermaphrodite flower is frequently seen in sex reverting hermaphrodite and male trees. It has a very short androperianth tube which is adnate to the base of the pistil, with five stamens on relatively long filaments. The lobes of the five carpellate pistils occupy positions which in the elongata flowers are occupied by stamens of the upper set. Superficially, the lobes of the carpels appear to lie opposite the corolla lobes rather than alternate as in the elongata flower. They are called pentandria because they contain only five stamens.

2.2.1 Description

A	Staminate ♂	:	Bracteolate, sessible in cluster or raceme, incomplete, actinomorphic, bracts-leafy or scaley, hypogynous, funnel shaped 2.5–3.2 cm long
	Calyx	:	Sepals 5, gamosepalous, small, light green, lobed 5
	Corolla	:	Petals 5, gamopetalous, tube-like, elongated, yellow-coloured
	Androecium	:	Stamens 10, in two whorls, interones are smaller, epipetalous (joined with petals), anthers-bilocular, introse
	Gynoecium	:	Absent
	Fruit	:	Absent
	Seeds	:	Absent
	Floral formula	:	♂ K (5) A5 + 5G0

(Continued)

B	Pistillate ♀	:	Borne on a raceme, corymb, sub-sessile, bracteolate
	Calyx	:	Petals 5, gamosepalous, light green
	Corolla	:	Petals 5, linear, deciduous, polypetalous, twisted aestivation
	Androecium	:	Absent
	Gynoecium	:	Sessile, style very short, stigma 5, dilated or linear, simple or lobed, ovary–superior, monocarpellary, parietal placentation, many seeded
	Fruit	:	Berry (pulpy in nature)
	Seeds	:	Blackish to brownish in colour, straight embryo, fleshy endosperm, cotyledons oblong and flat
	Floral formula	:	% ♀ K (5) C 5 A0 G1
C	Normal hermaphrodite ⚥	:	Long fruited type plants, flowers similar to pistillate type but the inflorescence is multi-flowered (5–6 flower corymb), corolla gamo, stamens 10 (5 + 5), sessile at base of petals, ovary usually functional
	Floral formula	:	% ♀ K (5) C (5) A 5 + 5 G1
D	Elongata hermaphrodite ⚥	:	Deformed/catface fruits, flowers similar to hermaphrodite type but five stamens (fused with carpels), sessile at base of petals, ovary usually functional
	Floral formula	:	% ♀ K (5) C (5) A 5 G1

The quality and quantity of pollen from three types of flowers, elongata, reduced elongata and staminate from the commercialised Thai papaya cultivar 'KhakNual', were determined using pollen morphology, pollen physical characters and pollen development processes. Pollen development progressed at the same pace in all three types of pollen-producing flowers and was consistent with pollen development in many angiosperms. Pollen morphology showed that papaya pollen grains are tricolporate, with three apertures, and there is no significant difference in diameter (25.18–25.72 μm) and weight (11.76–15.45 ng) among pollen sources. The staminate flower shows the lowest amount of pollen, with 12,368 pollen per anther, but higher viability and germination rates of 95.53% and 53.64%, respectively. In contrast, the elongata type shows the highest amount of pollen grains with 14,884 pollen per anther and the lowest viability and germination rates, 93.06% and 46.33%, respectively. The physical characteristics of pollen grains from reduced elongate and elongate flowers are similar. Reduced elongate flower type can donate pollen without self-pollinating (Phuangrat et al. 2013).

2.2.2 Flower Induction

Neither photoperiod nor temperature induces flowering in papaya. Each plant commences flowering only after a genetically determined number of nodes have developed following a short period of juvenility (Storey 1986). However, photoperiodic changes are responsible for sex reversals in certain phenotypically unstable forms of hermaphrodite and male trees. The reversal occurs at times closely related to the equinoxes and solstices (Storey 1986). Floral composition of sex reverting types may be influenced by temperature, water availability and other factors (Lange 1961;

Nakasone 1967; Storey 1969; Allan et al. 1987). Floral induction is also correlated with number of leaves produced. Under favourable conditions, a short-statured variety like ‘Betty’ produces its first inflorescence at the 24th node and a tall-statured variety such as ‘Solo’ not until the 48th node. The number of nodes and internodes often determine the height at which the first flowers appear. For variety ‘Betty’ it is about 40 cm and for ‘Solo’ about 140 cm (Storey 1986).

After papaya plants begin to flower, a flowering peduncle is produced in each leaf axis. Each peduncle produces several flowers of which one is terminal and two to five or more are in lateral positions. In female trees, it appears almost certain that only the terminal flower will set fruit with laterals dropping within a week after anthesis. Therefore, only one fruit per leaf axis is produced. On bisexual trees, it is also common for the terminal flower to set fruit with laterals dropping, thus, producing one fruit per panicle. However, one or two lateral flowers of some cultivars under favourable nutritional and moisture conditions tend to set fruits. In most instances, fruits of lateral flowers do not persist beyond two or three weeks.

2.2.3 Floral Biology and Pollination

For breeding purposes, terminal flowers are selected and tagged. They are emasculated before anthesis. While the peak time of anthesis is from 5.00 to 8.00 a.m. in all types of flowers, anther dehiscence begins about 6 h ahead of anthesis (Sharma and Bajpai 1969). Though no correlation has so far been reported between temperature and the rate of anthesis, the dehiscence, however, is increased markedly with rise in temperature and decline in the relative humidity. The pollens of hermaphrodite flowers obtained at the beginning of the flowering season show higher fertility (Sharma and Bajpai 1969). Five per cent sucrose solution gives best pollen germination under artificial conditions. Papaya pollens stored at room temperature in desiccators containing 64.8% H_2SO_4 and 10% relative humidity retain 50% viability up to 10 days. When stored in desiccators (63.15% H_2SO_4 and 10% RH) at lower temperature (0–4°C), the viability is 52.57% after 3 months of storage (Prakash and Dikshit 1963). The stigma becomes receptive 48 h before anthesis and continues to be receptive up to 72 h after anthesis. Usually for crossing, the flowers about to open are emasculated in the evening and bagged. They are pollinated in the morning next day with freshly collected or stored pollens and bagged again. Anthesis reaches a peak between 5.00 and 6.00 a.m. in flowers of all species studied, except pistillate flowers of *C. Cauliflora* (7.00–8.00 a.m.) and the staminate flowers of *C. Goudotina* (6.00–7.00 a.m.). Anthesis is sex dependent and stigma receptivity is maximum on the day of anthesis (Subramanyam and Iyer 1986).

It has often been observed that when flowers on female trees are pollinated with pollens obtained from the flowers of bisexual trees, they set maximum number of fruits. Plant breeders may encounter pollination problems with bisexual flowers due to their morphological differences. In most cultivars or breeding lines, the anthers extend directly over the stigmatic rays, which ensure automatic self-pollination. Flowers of this type may be bagged with assurance of self-pollination. In some lines and cultivars the position of the anthers, by virtue of having short filaments or adnation of the filaments at a flower position on the neck of the corolla tube, leaves considerable space between the stigmatic rays and the anthers. The anthers may be 5.0–8.0 mm below the stigma. In these cases, most of the bagged flowers fall. The problem may be alleviated by pollinating the flowers with pollen from the male flowers that are usually produced on andromonoecious (hermaphroditic) trees before bagging (Nakasone 1986). Studies on the Sunrise variety suggest occurrence of cleistogamy in this variety. The time from pollination to first ovule penetration was 25 h at 28°C temperature.

2.2.4 Pollinators

Papaya requires pollination for set fruit. The pollination of the dioecious flowers is primarily carried out by nocturnal moths known as hawk moths or sphinx moths (Martins and Johnson 2009). Although papaya flowers are visited by hawk moths, several beetles, skipper butterflies, bees, flies, and humming birds also visit papaya flowers (Figure 2.2). The actual species of hawk moths vary from site to site but, in general, any medium-to-large-bodied, relatively long-tongued species of hawk moth can serve as major floral visitors/pollinators, namely *Hippotion celerio*, *Herse convolvuli*, *Macroglossum trochilus*, *Daphnis nerii*, *Nephele comma* due to fast-flying and highly mobile insects, makes them efficient pollinators, whereas *Ceoliades* sp. (Hesperiidae) and *Sphingomorpha chlorea* (Noctuidae) are occasional visitors (Garrett 1995). Brown et al. (2012) reported that hawk moths are responsible for most of the pollen exchange. Moths visit the flowers searching for nectar. Pollen is transported on the moth's tongue by large species and rarely on the body by smaller species. Wind pollination has also been reported (Sritakae et al. 2011).

2.2.5 Parthenocarpy

Parthenocarpy in papaya is not a general rule but occurrences have been reported. There are apparently some female trees that exhibit parthenocarpy. In an experiment,

FIGURE 2.2 **(See colour insert.)** Major floral visitors/pollinators of papaya.

one set of flowers on a pistillate tree was bagged before anthesis and another set was pollinated before anthesis and then bagged. Both pollinated and unpollinated flowers produced fruits devoid of seeds. In still another case, bagged flowers produced seedless fruits, while unbagged flowers produced seeded fruits. Rodriquez-Pastor et al. (1990) have observed parthenocarpy at a rate of 35% in Sunrise and 5% in an advanced line 298F5. The parthenocarpic fruits are of adequate quality. Purohit (1980) also noted a tenfold increase in seed number of unbagged flowers from female trees with supplemental pollination. Under subtropical north Indian climate, controlled pollination done in August–September produces highest seed yield (Ram and Ray 1992).

2.2.6 Self-Incompatibility

Self-incompatibility in cultivars of *Carica papaya* seems to be rare as indicated by the lack of literature on this subject. However, Nakasone (1986) observed a hermaphrodite papaya strain to be partially self-incompatible as about 50% of the self-pollinated flowers dropped without fruit set. Some degree of self-incompatibility in the variety, Maradol from Cuba has also been reported as many self-pollinated flowers failed to set fruits. It is apparent that there are isolated cases exhibiting some degree of self-incompatibility. These cases are not the same as the anther/stigma–position relationship mentioned earlier (Ray 2002).

2.3 GENETICS

C. papaya with a somatic chromosome number, $2n = 2x = 18$, is the sole species of the genus *Carica* of Caricaceae, a family that includes six genera with at least 35 species (Storey 1976; Ram 1996; da Silva et al. 2007; Carvalho and Renner 2013). Caricaceae, consisting of four genera: (1) *Carica* with 22 species, (2) *Jacaratia* with six species, (3) *Jarilla* with one species and (4) *Cylicomorpha* with two species. The first three are indigenous to tropical America and the last one to equitorial Africa (Badillo 1971). Species having edible fruits are found only in *Carica*. Besides *C. papaya,* they are *C. chilensis, C. goudotiana, C. monoica* and *C. pubescens*. The fruit of *C. papaya* L. is generally eaten fresh. The fruits of other three species are eaten cooked as a vegetable or candied by cooking in syrup. In Peru, the leaves of *C. monoica* are cooked and eaten for greens.

2.3.1 Species

Among 22 *Carica* spp., following species are utilised in breeding programmes to induce resistance to diseases, waterlogged condition and frost.

C. candamarcensis Hook: It is the mountainous papaya and is small tree with cordate, palmately 5-lobed leaves and small yellow 5-angled 7.5–12.5 cm long fruit. The fruits are too acid to be used but can be stewed or made into jam.

C. erythrocarpa Linden and Andre: This is similar to *C. papaya* Linn. but the fruit has thin red flesh.

C. quercifolia Bant and Hook: The plant height reaches up to 2 m maximum and has oak-like leaves and cluster of small ellipsoid fruit 2.5–5 cm long with longitudinal strips that change from white to yellow when it ripens. It is harder than *C. candamarcensis* and the fruits though small are reported to contain a greater percentage of papain than *C. papaya.*

C. gracilis: It is a small slender ornamental species. It has compound leaves of five leaflets and each leaflet having wavy indentations, the middle leaflet being three.

C. monoica	:	(Desf): Monoecious plant, susceptible to virus.
C. microcapra	:	(Jacq): Dioecious plant, susceptible to virus.
C. cauliflora	:	(Jacq): Dioecious plant, resistance to virus (Figure 2.3).
C. goudotiana	:	(Solms-Lauback): Dioecious plant, susceptible to virus.
C. parviflora	:	(Solms): Dioecious plant, susceptible to virus
C. pennata	:	(Heilborn, Siensk): Dioecious plant, resistant to frost.
C. pubescens	:	(Lenne et Koch): Dioecious plant, resistant to distortion ringspot virus.
C. stipulate	:	(Badillo): Dioecious plant, resistant distortion ringspot virus.
C. horovitziana	:	(Badillo): Dioecious plant, susceptible to virus.
C. candicans	:	(Gray): Dioecious plant, resistant to distortion ringspot virus.
C. pentagona	:	(Heilorn): Dioecious plant resistant to frost.

C. cauliflora (Figure 2.3) is resistant to viruses, while *C. candamarcensis* and *C. pentagonia* are resistant to frost (Singh 1964). According to their crossibility, these species can be arranged in three groups; Group A: *C. monoica, C. microcarpa, C. cauliflora* and *C. candamarcensis*, Group B: *C. papaya* and Group C: *C. goudotiana.*

All the species in Group A are easily crossable with each other and produce viable seeds. Crosses between species in Group A and B do not form mature seed but immature

FIGURE 2.3 **(See colour insert.)** *V. cauliflora* tree in fruiting.

embryo in most of the cases can be cultivated through embryo culture. Crosses between species of Group B and C have not been successful. Species like *C. candamarcensis, C. guercifolia, C. monoica, C. cauliflora, C. goudotiana, C. petandra* and *C. microcarpa* have significance in breeding for biotic and abiotic stresses (Chadha 1992). *Carica pubescens, C. stipulata, C. Heibornii* and *C. candicans* have been reported to be resistant to 'Distortion RingSpot Virus' (Horovitz and Jimenez 1967).

The progeny of crosses with *V. caulifliora* were the only hybrids showing some virus resistance but they were unfruitful when attacked. There were no viable seeds and 30% of the fruits were seedless. *C. monoica* proved well adapted to Palmira, bore small, yellow fruits, but succumbed to virus. The introductions from Brazil were by far the most promising. 'Zapote' with rich, red flesh is much grown on the Atlantic coast of Colombia.

2.4 SEX FORMS

Except three species of *Carica*, all members of *Caricaceae* are dioecious. The three exceptions are *C. monoica, C. pubescens* and *C. papaya.* They have sexually ambivalent forms which go through 'sex reversals' in response to climatic and/or photoperiodic changes during the year. *C. monoica* is strictly monoecious but at certain times of the year may lack pistillate flowers. Plants of *C. pubescens* exist in three basic sex forms: pistillate, staminate and hermaphrodite. The pistillate and staminate plants are unresponsive to seasonal climatic changes. The hermaphrodite plants are sexually ambivalent, producing staminate, perfect and pistillate flowers in varying proportions at different times of the year. *C. papaya* exists in the same three basic sex forms as *C. pubescens.* The pistillate plant is stable. Staminate and hermaphrodite plants may be: (a) phenotypically stable or (b) phenotypically ambivalent, going through seasonal sex-reversal, during which they produce varying proportions of staminate, perfect and pistillate flowers.

2.4.1 Classification

Since papaya is a polygamous species, many forms of inflorescence have been reported. Generally, there are three types of flowers namely, staminate, pistillate and hermaphrodite but variation within each type has led to further classify them into six categories (Figure 2.4):

i. Dioecious pistillate – large sessile pistillate flowers in the leaf axil. It is the normal female, occurring in higher proportion than other forms.
ii. Dioecious staminate – long, narrow, tabular flowers on long peduncles hanging from leaf axil. It is described as the normal male.
iii. Andromonoecious – male and hermaphrodite flowers on a long peduncle.
iv. Polygamous – (a) male, female and hermaphrodite flowers on the same tree on long peduncles, (b) like (a) but hermaphrodite flowers being of two type namely, with 10 stamens and with 5 stamens. Fruits are long in shape.
v. Staminate – (a) staminate flowers on short peduncles directly, (b) hermaphrodite flowers with few female flowers on long peduncles and (c) separate staminate and cluster of staminate with hermaphrodite: (ci) hermaphrodite

FIGURE 2.4 **(See colour insert.)** Papaya (*C. papaya*) flowers: (a) male flowers, (b) male flower in longitudinal section, (c) female flower, (d) hermaphrodite flower, (e) hermaphrodite flower of pentandria type, (f) hermaphrodite flower with carpelloid stamens.

flowers on long peduncles, (cii) hermaphrodite flowers with few female flowers and (ciii) hermaphrodite flowers with many female flowers.

However, further studies have led to classify papaya flowers and its plants into groups:

2.4.2 Flower Types

Six types of flowers are known in papaya plant.

1. Typical female flower: It is a rather large flower of conical shape when closed, and when open, it has five petals spread from the base. The ovary is large with circular and smooth or slightly undulated. Fruits produced by these flowers are spherical or ovoid in shape (De Los Santos et al. 2000).
2. Pentandria: Similar to typical female flower when flower closed, but this type has five short anthers, which correspond in their orientation with the five petals that also spread from the base. The ovary has five deep longitudinal grooves that remain until maturity. Fruit develops a form from globular to egg shaped.
3. Hermaphrodite intermediate flower: The organisation is undefined; petals may be fused up to two-thirds of their length or free from the base. The number of anthers ranges from two to ten; the carpels range from five to

ten, with different degrees of fusion. This type of flower produces irregularly shaped fruit known as carpelodic (cat face), with little commercial value. These flowers appear more frequently when ambient temperatures are 24.5°C during the day and 15.5°C at night.

4. Hermaphrodite elongated flower: Petals of this type of flower are fused from one-fourth to three-fourths of their total length; ten anthers are observed, five long and five short. The ovary is long and when it contains five or more carpels, the form of the fruit varies from cylindrical to pear-shape (De Los Santos et al. 2000). Among different types of hermaphrodite flowers, this is the most commercially important.
5. Hermaphrodite sterile flower: It is a flower that resembles the former, but does not develop an ovary and hence it is sterile due to warm temperatures or water stress. It produces pollen only and considered a functional male flower (De Los Santos et al. 2000).
6. Typical male flower: This type of flower has a long and thin corolla containing anthers in two series of five, one series longer than the other. They have a rudimentary pistil no stigma and are non-functional (De Los Santos et al. 2000). According to aforementioned flower types, plants are also categorised into four groups:
   a. Group I – Pistillate or female plant producing Type A flowers.
   b. Group II – Hermaphrodite or bisexual plant may bear flowers of Type B, C and D. Mostly, it has Type B in summer, Type D in winter, Type C during transition periods and rarely Type E.
   c. Group III – Summer sterile hermaphrodite plant produces Type E in winter and pseudo Type D in summer (sometimes, it is considered as aberrant of Group II).
   d. Group IV – Staminate or male plant producing Type F flowers which are usually born but occasionally Type E flowers appear.

Storey (1958) classified papaya plants in 31 heritable phenotypes on the basis of peduncle length, ramification and seasonal sexual responses. Fifteen of these are variations among staminate plants and 15 are variations among hermaphrodite or andromonoecious plants. The remaining phenotype is the pistillate plant. Later, these phenotypic variations were broadly grouped into eight distinct types, namely (1) staminate, (2) teratological staminate, (3) reduced elongata, (4) elongata, (5) carpelloid elongata, (6) pentandria, (7), carpelloid and pentandria and (8) pistillate. This grouping greatly simplified numerous overlapping and removed confusions with respect to sex types. Fisher (1980) has reported that the perfect flower of the hermaphrodite, the pistillate flower of the female and the staminate flowers of the male are regarded as normal types. All polymorphic forms are regarded as teratological (i.e. abnormally modified), representing residual phylogenetic transitional stages between the perfect flower and the pistillate flower on the one hand, and the perfect flower and the staminate flower on the other hand (Ray 2002).

A hermaphrodite plant normally produces reduced elongata, elongata, carpelloid elongata, pentandria and carpelloid- and pentandria-type flowers (Figure 2.4). A male plant produces staminate flowers, while a female bears only pistillate flowers.

Teratological staminate flowers are produced by sex-reversing males. Depending upon the variations in environmental conditions, sex-reversing males can produce all the eight types of flowers. However, the hermaphrodite plants can have six types, except staminate and teratological staminate flowers. The female plants normally bear pistillate flowers, though rarely they can produce bisexual flowers (Ram 1996). Generally, the change in sex of male plants is towards hermaphroditism or femaleness. Hermaphrodite plants, however, do not produce staminate flowers. Similarly, female plants never undergo any change with regard to sex forms and, thus, never found to possess any hermaphrodite or male flowers. Femaleness is the most stable character, least affected by seasonal variations.

2.5 SEX IDENTIFICATION

Carica papaya is a polygamous diploid plant species with three basic sex types: male, female and hermaphrodite. Studies have been conducted for identifying the sex of papaya plant before flowering but till date no definite criteria have been found. Papaya seeds of deep brown colour produced a higher proportion of female plants than light coloured seeds. However, it was not clear whether this was due to relationship between seed colour and sex or due to relationship between age of seed at the time of picking, as indicated by colour and sex. He further observed that female seedlings at 12 weeks had higher stem elongation than male seedlings. Male plant had longer internodes and the female had greater stem girth in the flowering zone than male before flowering. Weaker seedlings had greater chances of being pistillate than strong ones. Staminate plants had noticeable lower mean height at first flower buds than the pistillate plants. Dark seeds of papaya were associated with high proportion of female and male plants, dark brown seeds of medium size were conducive to female flowers, whereas dark brown seeds of large size produced predominantly male plants. Identification of molecular and morphological markers like leaf markers and rate of growth at the juvenile and seedling stage for different sex types and their evaluation at flowering stage in five varieties of papaya namely, Ratna, Washington, Honeydew, CO-6 and CO-2, it was found that based on the leaf morphology and rate of growth, male and female seedlings could be identified at seedling stage. The study indicated that the seedlings start with single-lobed leaves and differentiate into three- and five-lobed leaves. Three-lobed leaves are predominant in males and five-lobed leaves in females. Leaf morphology and rate of growth at seedling stage have been exploited in identifying the presumptive male and female seedlings at the juvenile seedling stage (60–75 cm height). This is the first report of the identification of sex types at seedling stage and has great impact on economic returns to farmers and horticulturists. The earlier given results were confirmed through RAPD (Randomly Amplified Polymorphic DNA) for different sex types and at flowering stage in field (Reddy et al. 2012).

Chemical analysis to identify the sexes at seedling stage showed that when semen reagent (mercury dissolved in red fumic nitric acid and diluted) was added to alkaline water extract of dried leaves to which a saturated $CuSO_4$ solution was added, the resulting colour distinguished mature male and female plants. The same test on leaves of undifferentiated seedlings predicted sex correctly in 67% and 87% of male and female seedlings, respectively. Different protein patterns were observed in leaf and flower extracts of vegetative male and female plants in papaya. As the flowers developed some new bands were observed and some of the earlier ones disappeared and these changes were specific. In a chemical analysis study of papaya parts, it was concluded that it may be possible to predict the sex of papaya plants by a study of their physiological metabolism. Even the sex expression may be controlled by physiological adjustments. Papaya exhibits wide morphological and biological diversity of its types with prominent sex specific characters. The papaya plants can be either dioecious or gynodioecious with male and female parts befalling in the same plant. Monoecious, presence of male and female flowers on the same plant is also found in some related species of papaya. In papaya, the change of sex occurs in some plants at high temperature, where short stalked male flowers are produced instead of usual perfect flowers. Male or bisexual plants changing completely in female plants after being beheaded and some 'all male' plants occasionally producing small flowers with perfect pistils leading to abnormally slender fruits are also instances of change of sex in papaya. If the sex of dioecious papaya is identified at the seedling stage, prior to their transplantation in the field, cultivation of male and female plants in a desired ratio would be achieved and resources like planting space, fertilisers and water could be devoted to female plants. A single female papaya plant normally produces as many as 100 fruits in its life cycle and about 250 g of crude papain a year. Thus, an increase in the number of fruit-bearing plants per hectare of land would directly lead to increase in the yield of fruits and papain, making the cultivation more profitable. Therefore, it is of immense agricultural importance to identify the sex of papaya plants at the juvenile stage.

Rouging unwanted male plants from female and dioecious papaya plantings is a cumbersome procedure usually followed in papaya cultivation. Dioecious nature of papaya is one of the major constraints for papaya for large-scale cultivation. Over the past seven decades, various hypotheses, based on the knowledge and information available at the time, have been proposed to explain the genetics of the papaya's sex determination. A high-density genetic map of papaya was constructed using 54 F_2 plants derived from cultivars, Kapoho and SunUp with 1501 markers, including 1498 amplified fragment length polymorphism (AFLP) markers, the papaya ringspot virus coat protein marker, morphological sex type and fruit flesh colour. These markers were mapped into 12 linkage groups at a LOD score of 5.0 and recombination frequency of 0.25. The 12 major linkage groups covered a total length of 3294.2 cM, with an average distance of 2.2 cM between adjacent markers. This map revealed severe suppression of recombination around the sex determination locus with a total of 225 markers co-segregating with sex types. The cytosine bases were highly methylated in this region on the basis of the distribution of methylation-sensitive and insensitive markers. This high-density genetic map is essential for cloning of specific genes of interest such as the sex determination gene and for the integration of genetic and physical maps of papaya (Hao et al. 2004).

2.5.1 Sex Determination

Sex chromosomes are of great interest due to their role in sexual reproduction. Sex chromosomes evolve through the suppression of recombination in the sex determining region between the X and Y chromosomes which allows for each region to evolve and change independently. Papaya's trioecious sex is determined by a pair of incipient sex chromosomes. *Carica papaya* has been found to have nascent sex chromosomes (Liu et al. 2004). The X and Y chromosomes are cytologically homomorphic, but the Y chromosome has a recombination suppressed male-specific region (MSY) which has lost gene content, resulting in the lethal YY genotype. *Silene latifolia* has both a large male-specific non-recombining region on the Y chromosome, and pseudoautosomal region that do still recombine with the X chromosome (Scotti and Delph 2006; Delph et al. 2010). There are at least three regions in the male-specific region of the Y relevant to sex expression, one that suppresses femaleness and two that promote maleness (Zluvova et al. 2007). The sex chromosomes are heteromorphic, because the Y chromosome is larger than the X chromosome (Grabowska-Joachimiak and Joachimiak 2002).

Six stages of sex chromosome evolution have been proposed to explain the variation of the sex chromosomes in angiosperms (Figure 2.5) (Ming et al. 2011). In the first stage, a male-sterile mutation and a female-sterile mutation arise in close proximity on a chromosome, but recombination still occurs in this region. Strawberry is

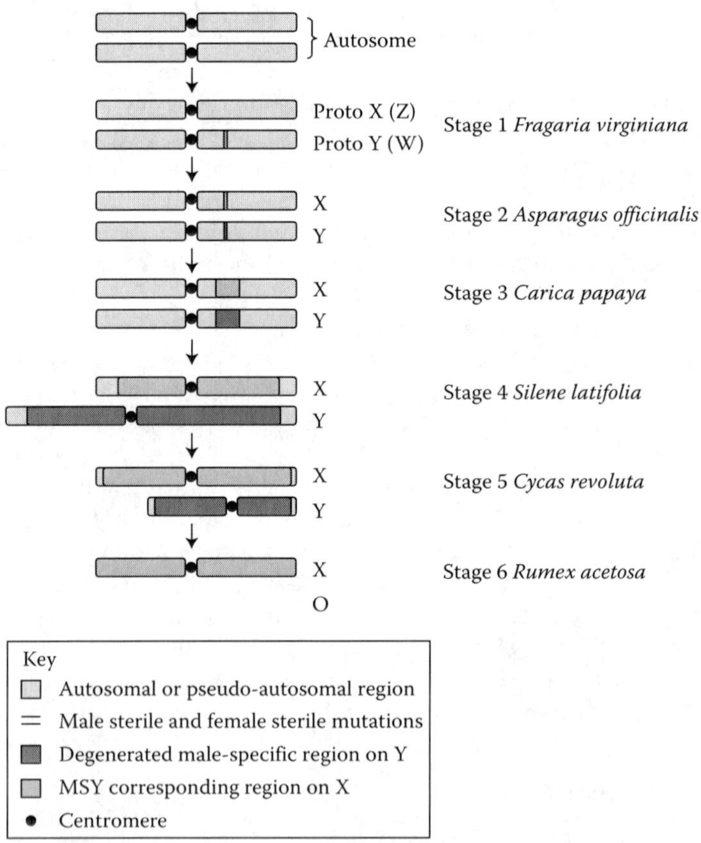

FIGURE 2.5 Six stages of sex chromosome evolution. (Adapted from Ming, R., A. Bendahmane and S. S. Renner, 2011, *Annual Review of Plant Biology*, 62: 485–514.)

an example of the earliest stage of sex chromosome evolution. In the second stage, recombination is suppressed at and around the mutations, a crucial step in sex chromosome evolution, leading to degeneration of the Y chromosome, though the YY genotype is still viable, as seen in asparagus. In the third stage, the suppression of recombination spreads to other loci, forming a male-specific region on the Y chromosome, and though the chromosomes appear to be homomorphic at this stage, some genes on the Y chromosome are lost through transposable element insertions, deleterious mutations and chromosomal rearrangements causing an inviable YY genotype. Papaya is an example of stage three. During the fourth stage, the MSY accumulates transposable elements and duplications, causing a DNA expansion. The non-recombining region spreads to majority of the Y chromosome and the sex chromosomes are heteromorphic, with the Y chromosome often becoming considerably larger than the X chromosome. Silene is a good example of an angiosperm in stage 4. During stage five, though a small portion of the sex chromosomes continues to recombine, keeping the pair together, severe degeneration of the Y chromosome occurs and many genes lose function, leading to the loss of the non-functional sequences, causing the Y chromosome to shrink. There are no current examples of angiosperm sex chromosomes in this stage. Finally, in stage six, the suppression of recombination spreads to the entire Y chromosome, causing the Y chromosome to be completely lost. Sex is then determined by an X to the autosome ratio, as is seen in Rumex.

Though many hypotheses had been made about the sex determination system of papaya, little concrete molecular data had been generated to verify which hypothesis accurately described what was occurring in papaya that led to these three sex types. Genetic cross data, phenotypic data and some early cytological observations were the only evidence used to form these early hypotheses. Scientists had no means to tackle the question of papaya sex determination until the applications of molecular techniques and biotechnology. The first method explored to detect papaya sex was through the use of sex-linked molecular markers. Microsatellite and sequence-characterised amplified region (SCAR) markers, which showed different banding patterns between the sex types, were successfully developed by different papaya research groups. This allows papaya sex to be determined during the vegetative state, but for commercial use, it is too costly to test thousands of seedlings and relocate them to the field (Parasnis et al. 2000; Deputy et al. 2002; Urasaki et al. 2002). Molecular markers were also used in producing multiple genetic linkage maps for papaya. The first map was constructed by Hofmeyr (1939), consisting of three morphological markers, namely sex, flower colour and stem colour. The second genetic linkage map consisted of 62 randomly amplified polymorphic DNA (RAPD) markers and mapped sex onto linkage Group 1 (Sondur et al. 1996). The third map incorporated 1498 amplified fragment length polymorphism (AFLP) markers, the papaya ringspot virus coat protein marker, sex and fruit flesh colour, totalling 1501 markers which were mapped onto 12 linkage groups (Ma et al. 2004). Most recently a high density genetic map using 712 simple sequence repeat (SSR) markers, designed from BAC end sequences, whole-genome shot gun sequences and a morphological marker resulted in nine large linkage groups and three small linkage groups (Chen et al. 2007). Sex was mapped onto linkage Group 1, one of the nine large linkage groups.

The construction of the papaya hermaphrodite BAC library, made up of 39,168 clones with an average insert size of 132 kb and 13.7X genome coverage, allowed for a new depth of exploration of the papaya sex chromosomes (Ming et al. 2001). The HSY was mapped with 225 sex co-segregating AFLP markers on linkage Group 1, showing severe suppressed recombination in this sex determining region. The SCAR markers were developed from sex co-segregating AFLP markers, and were used to screen the BAC library for physical mapping. Those BACs were extended by designing probes from their BAC end sequences to scan the BAC library for overlapping BACs. Through chromosome walking methods, a rudimentary physical map was produced with two major and three minor contigs that spanned 2.5 Mb. These efforts resulted in the discovery that severe suppression of recombination and degeneration is occurring in 10% of these homologous chromosomes, leading to the conclusion that these are in fact incipient sex chromosomes (Liu et al. 2004). Next, 50,661 BAC ends from 26,017 BAC clones were sequenced, allowing for chromosome walking techniques to be implicated to identify additional BACs in this area of interest (Lai et al. 2006). The completion of papaya whole genome sequence provided the resources to expedite the physical map construction for the hermaphrodite HSY and X-specific regions (Ming et al. 2010). The hermaphrodite BAC library clones were finger printed and BACs associated with the 2.5 Mb physical maps were used to discover contigs in the whole genome physical map that could aid in expanding the HSY through chromosome walking. Probes were designed from HSY BACS to detect corresponding X and male Y BACs to produce male MSY and the corresponding X-specific physical maps as well. Some of the BACs located on the HSY as well as a selection of paired X- and Y-specific BACs were sequenced and investigated. The HSY BACs showed a deficiency of genes, a large number of retro-elements, and gene duplication events compared to the X (Yu et al. 2007). Using genes found on both the HSY and X BACs, the divergence time of the X and Y^h chromosomes was estimated to be between 0.5 and 2.5 MYA, suggesting the sex chromosomes evolved at the genus or species level (Yu et al. 2008a). The Y and Y^h sequences were found to be nearly identical and likely arose from the same ancestral chromosome, instead of evolving separately. The divergence time between the Y and Y^h chromosomes was predicted to be 73,000 years (Yu et al. 2008b). In the male-specific regions of the compared BACs, various chromosomal rearrangements have occurred such as inversions, deletions, insertions, duplications and translocations. To date, the HSY and corresponding X region of the hermaphrodite have been physically mapped. Each physical map has only one remaining gap. The HSY physical map has a gap along Border-A which has been filled on the X physical map. The corresponding X region has a gap located towards the centre of the physical map between BACs 136D11 and 08K16, which is filled on the HSY physical map. The HSY spans ~8 Mb and the X spans ~5 Mb.

Earlier hypotheses reported a single gene with three alleles, a group of closely linked genes, genic balance of sex chromosome over autosomes, classical XY chromosomes and regulatory elements of the flower development pathway. Recent advancements in genomic technology make it possible to characterise the genomic region involved in sex determination at the molecular level. High density linkage mapping validated the hypothesis that predicted recombination suppression at the sex determination locus. Physical mapping and sample sequencing of the non-recombination region led to the conclusion that sex determination is controlled by a pair of primitive sex chromosomes with a small male-specific region (MSY) of the Y chromosome. It was reported that two sex determination genes control the sex determination pathway. One, a feminising or stamen suppressor gene causes stamen abortion before or at flower inception, while the other, a masculinising or carpel suppressor gene causes carpel abortion at a later flower developmental stage. Detailed physical mapping is beginning to reveal structural details about the sex determination region and sequencing is expected to uncover candidate sex determining genes. The cloning of sex determination genes and understanding the sex determination process could have profound application in papaya production (Ming et al. 2007).

Sex determination in papaya is controlled by a recently evolved XY chromosome pair, with two slightly different Y chromosomes controlling the development of males (Y) and hermaphrodites (Y^h). For the study of early sex chromosome evolution, Wang et al. (2012) sequenced the hermaphrodite-specific region of the Y^h chromosome (HSY) and its X counterpart, yielding an 8.1-megabase (Mb) HSY pseudomolecule and a 3.5-Mb sequence for the corresponding X region. The HSY is larger than the X region, mostly due to retrotransposon insertions. The papaya HSY differs from the X region by two large-scale inversions, the first of which likely caused the recombination suppression between the X and Y^h chromosomes, followed by numerous additional chromosomal rearrangements. Altogether, including the X and/or HSY regions, 124 transcription units were annotated, including 50 functional pairs present in both the X and HSY. Ten HSY genes had functional homologs elsewhere in the papaya autosomal regions, suggesting movement of genes onto the HSY, whereas the X region had none. Sequence divergence between 70 transcripts shared by the X and HSY revealed two evolutionary strata in the X chromosome, corresponding to the two inversions on the HSY, the older of which evolved about 7.0 million years ago. Gene content differences between the HSY and X are greatest in the older stratum, whereas the gene content and order of the collinear regions are identical (Wang et al. 2012).

2.5.2 Cytogenetics of Sex Chromosomes

Papaya has nine chromosome pairs, in which seven are metacentric and the remaining two pairs are sub-metacentric (Ming et al. 2008). Papaya chromosomes are small and uniform in morphology, making them hard to differentiate using length, arm ratios, or banding patterns, which led to difficulties in identifying sex chromosomes in early studies of papaya sex determination (Wai et al. 2010). In an early investigation, precocious separation was observed between a chromosome pair during anaphase I of meiosis of a pollen mother cell in males and hermaphrodites. Recombination occurs throughout the homologous regions of the sex chromosomes, but it is suppressed in the sex specific region of the chromosome pairs (~13% of the sex chromosomes) (Zhang et al. 2008). Recombination rate recovers and elevates in the border regions of the HSY (Yu et al. 2009). To link chromosomes to their genetic sequences, genetic mapping of the papaya genome was carried out and resulted in 12 genetic linkage groups, including nine major and three minor linkage groups (Chen et al. 2007). The three minor linkage groups 10, 11 and 12 were merged with major linkage groups 8, 9 and 7, respectively, using molecular cytogenetic approaches (Wai et al. 2010). To determine which chromosome corresponds to which genetic linkage group, linkage group-specific BACs were used as probes for fluorescence *in situ* hybridisation (FISH) in papaya meiotic pachytene chromosomes (Zhang et al. 2010). The X and Y chromosomes were identified as the second longest chromosome pair and were designated as chromosome 1 in the karyotype. The remaining papaya chromosomes were numbered according to length, chromosome 2 being the longest and chromosome 9 being the shortest. To locate the HSY region on the sex chromosomes, two confirmed BACs in this area were directly hybridised to interphase, prometaphase, metaphase and anaphase chromosomes (Yu et al. 2007). These BACs hybridised near the centromere of the Y^h. One BAC had a weaker signal on the X, which suggested the sequences on the HSY and X in this region were still relatively conserved. Since the second BAC only hybridised to the Y^h, it was likely that BAC sequence had diverged considerably. Pachytene FISH was also utilised to map one of these HSY-specific BACs, along with its neighbouring BACs and a non-HSY BAC. The HSY BACs showed strong signals only on the Y^h, whereas the non-HSY BAC showed a strong signal on a different homologous pair of chromosomes. This study verified the identity of the X and Y^h chromosomes and located the HSY near the centromere on the Y^h. To further explore the structure of the MSY and X regions, hermaphrodite meiotic pachytene chromosomes were stained with 4,6-diamidino-2-phenylindole (DAPI), which stains heterochromatic regions of chromosomes (Zhang et al. 2008). Based on the staining, the arms of the chromosomes were mostly euchromatic, but clusters of heterochromatin were found around the centromere. Specifically, the XY^h bivalent was mostly euchromatic, with the X chromosome being the most euchromatic chromosome in the papaya genome, but five knob-like regions of heterochromatin, numbered K1 through K5, were found in the HSY (Zhang et al. 2010). The largest knob, K1, was shared between the HSY and the X, but K_2, K_3, K_4 and K_5 were only found on the HSY. The knobs were also found to be highly methylated. The heterochromatic knobs were likely the result of transposable elements and the high DNA methylation in these regions could be a defence mechanism against transposable element invasion. The HSY contained two small regions of 5S rDNA, which is an element of the large subunit of the ribosome involved in translation (Zhang et al. 2010). These regions were associated with K2 and K4. The X chromosome did not exhibit 5S rDNA. The accumulation of 5S rDNA in the HSY likely led to the materialisation of heterochromatin and assisted in the differentiation of the sex chromosomes (Zhang et al. 2010). During X and Y^h chromosomal pairing, a slight curve in the Y^h chromosome occurred to allow for pairing (Zhang et al. 2008). The region around K4 had accumulated considerably more DNA then its X corresponding region, causing the curving of the Y^h chromosome during pairing. By implementing meiotic metaphase I-based FISH using Y^h-specific BACs, the centromere of the Y^h chromosome was found to be located in the HSY, specifically associating the centromere with K4 (Zhang et al. 2008). The area around K4 showed more sequence divergence from the X than other regions of the HSY.

2.6 INHERITANCE PATTERN

Sex in papaya is controlled by five pairs of genes which occur in three sex determining complexes in sixth chromosome. Because of tight linkage between the genes, the sex-determining complexes produce phenotypic results analogous to that produced by three alleles of a single gene with pleiotropic effects. For practical convenience, the complexes are generally designated as: M_1 dominant for maleness, M_2 dominant for hermophroditism and *m* recessive for femaleness. The combination M_1M_1, M_2M_1 and M_2M_2 are lethal and, thus, fail to produce viable seeds. M_1m gives male trees, M_2m hermaphrodite trees and *mm* female trees. The crossing of two plants differing in sexform produces either two (male or female) or three (male, female and hermaphrodite plants) in set ratios (Table 2.1). Yadav and Prasad (1990) observed some variation in the ratio with hybrids from dioecious × gynodioecious crosses. Such a cross produced the ratio of 1 male: 2 female: 1 hermaphrodite, with the exception of the two crosses where female and hermaphrodite progenies were produced in the ratio of 1:1 and 3:1. These observations confirm that the inheritance of sex is under polygenic control.

The size and shape of the fruits are determined by the parentage, particularly the type of flowers used in pollination. By selfing hermaphrodite of known genetical constitution, for example 'Solo', uniform pyriform fruits are produced on the hermaphrodite progeny and uniform larger round fruits on the female progeny. By crossing mmxM_2m, half the progeny will be female with round fruits and half will be hermaphrodite with cylindrical fruits. Thus, all the progeny will be with fruit. However, one of the difficulties of the breeder is that there are no reliable characters to distinguish male, hermaphrodite and female trees until they flower. In dioecious varieties, both male and female trees should be progeny of the same parents which have the desired characters. Maintaining characters in hermaphrodite form is easier. Also, genetic manipulation of hermaphrodite plants is easier by conventional breeding methods as compared to other sex forms.

TABLE 2.1
Crossing among the Basic Sex Forms in Papaya and Ratio of Segregating Progenies

Cross	Female (mm)	Progenies: Hermaphrodite (M_2m)	Male (M_1m)	Non-Viable (Lethal)
F* × S (mm × M_1m)	1	–	1	–
F × H (mm × M_2m)	1	1	–	–
H × H (M_2m × M_2m)	1	2	–	1 (M_2M_2)
S × S (M_1m × M_1m)	1	–	2	1 (M_1M_1)
H × S (M_2m × M_1m)	1	1	1	1 (M_2M_1)
S × H (M_1m × M_2m)	1	1	1	1 (M_1M_2)

Source: Adapted from Hofmeyr, J. D. J, 1967, *Agronomy Tropical,* 17:345–51.
F* = Female; H = Hermaphrodite; S = Staminate.

Different hypotheses have been put forward on the genetics of sex determination in papaya. Hofmeyr's (Hofmeyr 1967) hypothesis involves genic balance. The symbols M_1 and M_2 represent inert or inactivated regions of slightly different lengths on sex chromosomes from which vital genes are missing. This accounts for zygotic lethality of M_1M_1, M_2M_2 and M_1M_2 genotypes (Table 2.1).

The greater concentration of genes for femaleness is on the sex chromosomes, whereas that for maleness on the autosomes. Thus, the genotype *mm* is pistillate and its homozygosity confers phenotypic stability. Since M_1 is longer of the inert regions, it is expressed phenotypically as staminate because of the greater influence of the autosomal factors. The shorter M_2 region is less influenced by autosomal genes, so, the M_2m genotype is expressed phenotypically as hermaphrodite or andromonoecious. The heterozygosity of M_1m and M_2m renders them susceptible to alternation in phenotypic expression by external influences (Ray 2002).

The hypothesis of Horovitz and Jimenez (1967) proposes a reversionary process for sex determination and expression. Its basic assumption is that dioecism is the primitive state in *Caricaceae* and sex determination is of the classical XX–XY type with heterogametic male and YY lethal to the zygote. At some point in time, a sexual ambivalent form occurred in the genus from which three present day exceptions to dioecism (*C. monoica, C. pubescens* and *C. papaya*) arose. In *C. papaya*, the ambisexual mechanism built up on the Y chromosomes, giving rise to a modified homologue, the Y_2 chromosome which (in the heterogametic genotype XY_2) is expressed as the sexually ambivalent andromonoecious form. This occurred without alteration of the × chromosome, which explains the stability of the pistillate form. This hypothesis holds, therefore, that andromonoecism and polygamy followed the evolution of XX–YY system and are of fairly recent origin.

Storey (1967, 1969) hypothesised progressive evolution of dioecism in the family. Sexual differentiation in the form of dioecism followed the derivation of unisexual flowers. Since dioecism seems to be the evolutionary norm in *Caricaceae*, it is possible that ambisexual forms owe their continued existence to human selection. In Storey's (Storey 1976) revised hypothesis, the symbol (SA) represents the sum of the factors involved in transmuting the ancestral androecism into the present day gynoecium; (SA) represents normal androecium development; (SG) represents the factor or factors responsible for suppression of the gynoecium in the staminate flower; (SG) permits the (SA) factors to function ontogenetically in developing the replacement gynoecium. The symbol *'l'* represents the recessive sex-linked zygotic lethal factor that enforces heterozygosity on the staminate and andromonoecious plants; *C* represents the factor that prevents crossing-over between the sex-determining factors and the lethal factor, accounting for the non-existence of pistillate plants carrying the *'l'* factor. As in other two hypotheses, heterozygosity permits ambisexuality. The sex determining genotypes are expressed as follows:

i. Staminate and andromonoecious: (sa)*l* C (SG)(SA) ++ (sg)
ii. Pistillate: (SA) ++ (sg) (SA) ++ (sg)

For convenience, the sex homologues may also be represented as: M^h – andromonoecious or hermaphrodite; M^s – androecious, that is staminate or male; m – gynoecious, that is, pistillate or female. Length and ramification of the inflorescences are secondary sex characters. The alleles that give them expression are on the sex chromosomes but linkage with the sex-determining factors is not absolute. In effect, all staminate and andromonoecious forms are identical. Therefore, the classification of plants as one form or other depends largely upon the nature of the inflorescence. Elucidating the structure of sex chromosomes, Ram et al. (1985a, b) have postulated that multiple allelic genes are at play in determining sex of papaya plants. They suggested separate symbols for pure male ($M_1^{rr}m$) and sex reversing males ($M_1^{Rr}m$) recognising dominant gene for homozygous sex reversing maleness as M_1^{RR}; for heterozygous sex reversing maleness as M_1^{Rr} and for pure maleness as M_1^{rr}. Chromosomal structures proposed by Ram et al. (1985a, b) for different sexes are as follows:

Female	mp	m	V	suF	X
	mp	m	v	suF	X
Hermaphrodite	mp	m	V	suF	X
	mp	M_2	v	suF	Y
Pure Male	mp	m	V	suF	X
	Mp	M_1^{rr}	v	suF	Y
Sex reversing male	mp	m	V	suF	X
	Mp	M_1^{Rr}	v	suF	Y

In this hypothesis, *Mp* or *mp* is the main distinguishing factor between male and hermaphrodite plants (MP stands for male plant and *mp* for hermaphrodite plant). It has been assumed that X chromosome carries a vitality gene (V) which is recessive (v) in Y chromosome. The homozygous recessive (vv) state is lethal; thus, homozygous recessive individuals are never seen in segregating populations; only the heterozygote (Vv) and homozygous dominants (VV) can grow. The presence of the lethal gene in recessive state in both males and hermaphrodites adds heterozygosity to the sex forms. A dominant gene for suppressing femaleness (SuF) is present only in a pure male. As a result, this plant never produces any fruit. The suppressor gene (SuF) is absent in sex reversing males. Thus, all such plants are capable to bear fruits. This explanation is a partial modification to the Hofmeyr's (Hofmeyr 1967) hypothesis. However, there still exists an urgent need to investigate the order of the genes in the sex determining segment of a chromosome. Only detailed investigations will reveal whether the above hypotheses are relevant or speculative.

Detailed studies on genetic and breeding aspects of the crop (Hofmeyr 1938, 1942, 1967; Nakasone and Storey 1955; Storey 1953; Yadav and Prasad 1990) have revealed that yellow flower colour is dominant to white, purple stem colour is dominant to red, yellow flesh colour is dominant to red, dwarfness in height is recessive to normal tall stature, diminutive plant (short slender trunk, small leaves) is recessive to large plant, crimpled leaf is recessive to normal leaf, rugose leaf recessive to normal smooth leaf, waxy leaf is recessive to normal flat-bladed leaf and grey seed-coat colour is dominant to black seed-coat colour. Morshidi (1998) studying the inheritance of isozymes in papaya, has observed Mendelian inheritance pattern for eight out of nine polymorphic loci. The highest number of fruits per plant and the highest number of seeds per fruit with the lowest fruit length are the most important selection indices to identify a plant type producing maximum fruit yield in papaya (Dwivedi et al. 1998).

2.7 BREEDING

2.7.1 Objectives

The main breeding objective is to develop cultivars with higher yield of better quality fruits. Six principal characters namely, yield, colour and texture of the flesh, fruit size, sweetness and storage are considered the most desirable traits in all breeding endeavours. Papaya plants suffer greatly from several viral diseases. Resistance to the virus infection would indeed be the first priority in breeding. Some other problems that need to be solved to enhance the probability of increased fruit and latex production are

- Elimination of ambisexual andromonoecious or hermaphrodite forms that tend to become female sterile at certain time of the year or show a tendency towards stamen carpelloidy.
- Development of homozygous hermaphrodite forms by possible elimination of the zygotic lethal factor or by *in vitro* culture of embryo.
- Inducing a sex-linked vegetative character for eliminating unwanted sex forms in the early seedling stage.
- Early and low bearing with short internodes.
- Selection for inflorescences of moderate length (about 7.5–10.0 cm) bearing single fruit, to rule out crowding with resulting misshapen fruits.
- Selection for an ovarian cavity circular in transverse section, with an easily separable placenta.
- Uniformity of fruit shape, texture and flavour for export trade.
- Breeding of dioecious cultivars for regions where andromonoecious types are excessively sexually ambivalent.
- Breeding to expand regions of production by hybridising with species that are more cold tolerant (e.g. *C. pubescens*) or tolerant to excessive soil moisture or salinity.
- Breeding for increased latex yield for papain production.

2.7.2 Methods

2.7.2.1 Backcross Breeding

Backcross breeding method has been frequently used in papaya improvement programmes. Its objective is to add easily heritable desirable alleles from a non-recurrent parent to the genetic background of a recurrent parent. Since papaya is a highly cross-pollinated crop, a great deal of variation exists in shape, size, quality, taste, colour and flavour of the fruit. With backcrossing, increased homozygosity for recurrent parent alleles will occur leading to fair uniformity in desirable traits. The rate with which this takes place for each locus can be calculated by the formula $(2^b - 1)/2^b$ where b is the number of backcrosses. For example, in the BC_6 (the sixth generation of backcrossing) 127/128 or 99.22% of the segregating loci for a particular trait will be homozygous. The outstanding dioecious papaya genotypes like, Pusa Dwarf and Hortus Gold, rapidly degenerate due to out-crossing. Backcrossing is often used for the recovery of desirable traits in such cultivars. Ram (1993) described the use of backcrossing in the improvement of inbred lines for use in developing papaya hybrids. For this, suitable male plants are selected from the same progeny which have resemblance to female plants in vegetative characters, such as stem and leaf colour, stem thickness and height at flowering. The typical female plant having maximum resemblance to a cultivar should be crossed with such three male plants identified earlier for the purpose. After confirmation, the best male plant should be retained for maximum possible generations for backcrossing. Due to dioecism in papaya, it may take 18 generations (36 years) to become almost completely homozygous (Hofmeyr 1953). Backcrossing with the retained male plant automatically reduces this period as it is difficult to retain the plant for such a long period in the field. Progenies rose from BC_1 or BC_2 inbred lines are screened and desirable female plants are selected for further sib-mating. The process is to be continued for 7–8 generations to achieve uniformity for a particular character or a group of characters. In backcrossing of a dioecious strain, seedlings segregate in 1 female: 1 male ratio (i.e. in equal proportion).

Although backcrossing does have some specific advantages as a breeding technique, a serious weakness called inbreeding depression is an undesirable fall-out of its use. The magnitude of inbreeding depression (i.e. decrease in vigour) is not the same in all lines produced by inbreeding. The point, after several generations of inbreeding, at which no further decrease of fitness and vigour occurs, is referred to as the 'inbreeding minimum'. The crossing of inbred lines which have reached their inbreeding minimum frequently results in heterosis. The cause of inbreeding depression is associated with the uncovering of deleterious recessives and lethals in homozygous genotypes. However, Hamilton's (1954) study in inbreeding and crossbreeding of two Solo strains showed that inbreeding has no ill effects on plant vigour.

Breeding for gynodioecious lines involves selfing of a regular and prolific bearing hermaphrodite or sibmating the female with the hermaphrodite. Suitable hermaphrodite plants whose sex does not change with climatic variations are selected. As done in the case of dioecious lines, here also the desirable female is crossed with at least three hermaphrodite plants as male parent. Sib-mating in female and selfing in hermaphrodite may go simultaneously to speed up the breeding programme of the various types of flowers produced by a hermaphrodite plant; elongata and pentandria types are selected for selfing. This process is to be continued for 7–8 generations for satisfactory homozygosity to be achieved. In this method, the sex ratio (female: hermaphrodite) is found to be 1:1 in sib-mating and 1:2 in selfing of hermaphrodites (Ram 1996). The major advantage of developing a gynodioecious line is that all plants bear fruits. The time consumed in attaining complete homozygosity through backcross breeding is reduced substantially if anther culture is adopted.

2.7.2.2 Recurrent Selection

The basic technique in recurrent selection is the identification of individuals with superior genotypes, and their subsequent inter-mating to produce a new population. Some type of progeny test, depending on the selection system, may be necessary to measure the genetic worth of the parents. Parental genotypes are retained often by controlled selfing, so that by following the process of selection and progeny evaluation, the best one can be grown and inter-crossed. After the inter-crossing of superior plants, selection can again be practiced in the new population. The recurring population improvement concept leads to the name 'recurrent' selection. This can be used for selecting superior types in dioecious as well as gynodioecious lines. Smith (1970) has amply emphasised the role of selection and controlled pollination in papaya. According to Dwivedi (1998) indirect selection of plants on the basis of higher number and weight of seeds per fruit, higher number of fruits and plant and low to moderate fruit length is more effective in increasing fruit yield than direct selection. In addition, selection of fruits for higher peel weight, fruit weight and diameter is expected to improve pulp weight in papaya (Dwivedi and Jha 1999).

2.7.3 Hybrid Breeding

2.7.3.1 Heterosis

Exploitation of heterosis, that is, producing F_1 progenies superior to the better parent involved in a cross has ample prospect in papaya improvement. Dai (1960) reported heterosis in the cross between 'Philippine' and 'Solo' varieties. The F_1 exhibited reduced number of seeds and enhanced vigour. A high positive heterosis for fruit size and number of seeds was observed by Sah and Shanmugavelu (1975) in the cross, CO-1 × Coorg Honeydew at Coimbatore. Iyer and Subramanyam (1981) observed heterosis for the vegetative characters, fruit yield and its components. Heterosis up to 11.4% for yield was obtained in the cross, Solo Yellow × Washington. High relative heterosis for fruits in certain crosses was observed in the combination, Solo Yellow Sweet × Washington; Pink Flesh Sweet × Coorg Honey; Pink Flesh Sweet × Washington and Thailand × Washington. High heterosis for potential economic competitiveness was noticed in Thailand × Washington. In crosses between Pusa Delicious × Halflong and Pusa Delicious × Homestead, high heterotic responses (47.34% and 39.77%, respectively) were observed with respect to fruit yield over mid-parent. Similarly, enough heterosis (31.29%) existed in the progeny obtained from Pusa Delicious × Homestead with respect to fruit yield over better parent. As regards fruit length, crossing between Washington × Ramnagar Local showed better heterosis, that is, 44.63% over mid-parent and 36.92% over better parent (Ram, 1982). Ram et al. (1999) have reported positive heterosis for fruit yield and number of fruits per plant, however, the heterosis was negative for characters like, single fruit weight (average), fruiting length, first fruiting node, plant girth, plant height and days to first flowering. Dwivedi (1998) has also indicated a negative influence of fruit length on fruit yield and pulp weight.

2.7.3.2 Combining Ability

In order to make hybrids a reality, identification of highly productive heterotic combinations is very important. It might be logical to assume that the greatest heterosis would be displayed by crosses of pure inbreds which have undergone maximum inbreeding depression. Heterosis dependent on combining ability can occur in any combination of pure inbreds.

General combining ability (GCA) is expressed in the progeny of an inbred crossed with many genotypes and is primarily the result of additive gene action. To test for GCA, five papaya inbreds namely, Pusa Delicious, Pusa Majesty, Washington, Sunrise Solo and Waimanalo were crossed with three tester(s), namely Homestead, Halflong and Ramnagar Local. The best GCA was found in Pusa Delicious and Ramnagar Local was the best tester (Ram 1982). This study indicated that higher performance needs to be correlated with high GCA effects.

Specific combining ability (SCA) is important in the identification of valuable inbred lines for use in hybrid production. It is the expression of performance between any two inbred lines and is attributed to dominant, epistatic and additive gene action. Ram (1982) made several crosses and observed that Washington × Ramnagar Local had the highest SCA for the fruit yield per plant, followed by Pusa Delicious × Homestead, Waimanalo × Halflong and Pusa Delicious × Halflong. It was observed that the best general combining line and tester did not necessarily posses/show the maximum SCA values. It was the combination of average combiners from both the inbred lines on one hand and testers on the other which showed the maximum SCA (Ram 1993). Dinesh et al. (1991) observed that variety, Thailand was a good combiner for fruit length, breadth and volume, while Sunrise Solo and Waimanalo were good combiners for quality characters.

The selection of resistant genotypes is a sustainable alternative to disease control in papaya cultivation. However, total resistance has not been observed in commercial papaya genotypes with some degree of selection. There is, however, the possibility that crosses generate hybrids with higher resistance levels to fungal diseases. Thus, crop breeding can contribute to the selection of resistant genotypes and/or indicated the best hybrid combinations, based on the hybrid vigour also known as heterosis. The hybrid vigour is directly related to the degree of genetic divergence of parents involved. However, a high genetic divergence does not necessarily increase the expression of heterosis. It is, therefore, important to use methods that identify the best combinations (Duarte et al. 2003). Vivas et al. (2011) evaluated testers to estimate combining ability and select hybrids resistant to black spot, phoma spot and chocolate spot. The severity of phoma spot and black spot on leaves, and the lesion area of black spot and chocolate spot on fruits were evaluated in two seasons. The combining ability of crosses is negative for all traits: tester 'JS 12' with 'Sunrise Solo' and 'Kaphoro SoloPV'; tester 'Americano' with 'Caliman M_5', 'Sunrise Solo', 'Baixinho de Santa Amalia' and 'Waimanalo'; and tester 'Maradol' with 'Caliman G', 'Caliman a.m.' and 'Sunrise Solo PT'. These results may be useful in breeding for disease resistance by hybridisation.

2.7.3.3 Hybrids

Although some hybrids have been developed following inter-varietal or inter-generic crosses, there still exists great scope for development of superior hybrids for better yield and quality. This is more relevant particularly for the production of F_1 hybrid seeds. At present, no private agency is producing F_1 papaya seeds in India. Very limited quantity is available at research stations and that too for experimental needs only. Thus, there is an urgent need to produce large quantity of F_1 papaya seeds. Besides this, two very vigorous F_1s from specific inbred combinations can also be crossed together to produce the hybrid seeds to meet the large demand. Because of heterosis associated with each F_1, the quantity of hybrid seed production can, thus, be raised.

But such seeds involve the risk of higher segregating population and sometimes may not be *at par* with F_1 in performance for yield and quality. Nevertheless, one can expect substantial yield gain with hybrid over the conventional variety. Cariflora derived by crossing two dioecious lines, K_2 and K_3 ($K_2 \times K_3$). It bears round fruits (13.5–14.5 cm diameter) with sweet yellowish flesh, good aroma and good quality. The ratio of male: female flowers or plant is 1:1. It is dioecious and highly tolerant to PRSV. Two hybrids, IIHR-39 (Sunrise Solo × Pink Flesh Sweet) and IIHR-54 (Waimanalo × Pink Flesh Sweet) developed at Indian Institute of Horticultural Research, Bangalore have been found highly promising on account of superior fruit quality. They bear medium-sized fruits, sweet in taste with good shelf life. Both are gynodioecious. IIHR-39 has now been released as Surya (Dinesh and Yadav 1998). Another promising hybrid, HPSC-3 (Tripura Local × Honeydew) has been developed at ICAR Research Complex, Tripura. It possesses high yield potential and resistance to mosaic virus (Singh and Sharma 1996).

2.7.4 Inter-Specific Hybridisation

The brief account of crossing relationships and inherent resistance of *C. monoica* to virus disease, namely bunchy top, has involved an intensive search of some of the other species to seek breeding stock which might carry immunity and higher resistance and to attempt the incorporation of such immunity into the ordinary papaya which is highly susceptible. A wide range of inter-specific and reciprocal combinations has been made by utilising some of the lesser known species. *Carica cauliflora* has been found to be resistant to mosaic virus (Capoor and Verma 1961). Horovitz and Jimenez (1967) have reported that *C. cauliflora, C. pubescence, C. stipulate* and *C. candicans* are resistant to distortion ringspot virus. Zerpa (1959) made several inter-specific crosses, namely *C. cauliflora*x *C. microcarpa, C. monoica* × *C. cauliflora, C. monoica* × *C. candamarcensis,* F_1 *(C. monoica* × *C. cauliflora)* × *C. candamarcensis* and *C. cauliflora* × *C. candamarcensis.* Cytological investigations of these crosses revealed that only the cross, *C. cauliflora* × *C. candamarcensis* produces multivalents at metaphase I, while others could produce bivalents indicating high inter-specific genetic affinity. Mekako and Nakasone (1975) while making inter-specific hybrids using six *Carica* species obtained viable seeds from *C. monoica* × *C. goudotina, C. parviflora* × *C. goudotiana, C. goudotiana* × *C. monoica* and *C. cauliflora* × *C. penata.* They also observed heterosis for vegetative and fruit yield characters in the crosses, *C. parviflora* × *C. goudotiana* and *C. cauliflora* × *C. monoica.* Subramanyam and Iyer (1981) and Iyer and Subramanyam (1984) attempted inter-specific hybridisation and found crossability barrier operating in many species. Their study showed that F_1 hybrid of *C. cauliflora* × *C. monoica* when crossed with *C. papaya* resulted in fertile hybrids. Although *C. cauliflora* and *C. monoica* are incompatible with *C. papaya*, hybrids of *C. cauliflora* × *C. monoica* are compatible. Iyer et al. (1987) later reported that inter-specific hybridisation between *C. papaya* and *C. cauliflora* easily produced F_1 plants. After backcrossing with *C. papaya,* they evolved a line, 21–19 showing resistance to mosaic virus with normal fruit quality. Magdalita et al. (1997) found *C. papaya* × *Carica cauliflora* hybrids resistant to Australian papaya ringspot potyvirus type P (PRSV-P) isolates.

Drew et al. (1998) developed the protocols involving biotechnological skills for hybridisation of papaya with related *Carica* species that are resistant to PRSV-P (*C. cauliflora, C. quercifolia* and *C. pubescens*). A highly efficient protocol was used for rescue and germination of *C. papaya* × *C. cauliflora* immature embryos. Embryos were made to germinate from embryogenic cultures on hormone-free agar solidified medium and multiple hybrid plants were produced. The *C. papaya* × *C. cauliflora* hybrids lacked vigour and were generally, infertile. Subsequently, the protocol was adopted to produce hybrids between *C. papaya* and other PRSV-P resistant species, *C. quercifolia* and *C. pubescens*. Hybrid plants grew vigorously in the field and a few *C. papaya* × *C. quercifolia* plants produced some viable pollens. Inter-specific hybrid plants between *C. papaya, C. goudotiana* and *C. parviflora* were also produced. *C. parviflora* was hybridised with *C. pubescens* and *C. gaudotiana*. Plants of all these crosses could be grown in glasshouse or in field. It is envisaged that these procedures may also allow access to other characteristics of wild species, such as *Phytophthora palmivora* resistance (*C. gaudotiana*), high sugar content (*C. quercifolia*) and cold tolerance (*C. pubescence*) (Drew et al. 1998). Magdalita et al. (1998) have also described an efficient protocol for inter-specific hybridisation between *C. papaya* L. × *C. cauliflora* Jacq. that involved the use of highly viable pollen of *C. cauliflora* produced during summer, autumn and spring; the use of an isolation time ranging from 90–120 days post pollination of hybrid embryos and the use of most compatible *C. papaya* cv., 2001 for crossing with *C. cauliflora*. The hybrids developed this way are reported to be resistant to PRSV-P (Ray 2002).

2.7.5 Polyploidy

A diploid commercial dwarf cultivar, Wonder Blight and 40 anther-derived papaya strains were raised in the same greenhouse. Morphological data based on sizes of stems, leaves, flowers, fruits, parthenocarpic ability and fruit yield were also collected. The anther-derived papaya strains turned out to be all female, but were variable in ploidy and morphology. The anther-derived plants were of different ploidy, namely haploids, diploids, triploids and tetraploids. Morphologically, even plants of the same ploidy were variable in height, parthenocarpic ability, fruit size, shape and yield. To conclude, the female papaya plants derived from the male gametophyte originated from the microspores. The haploids and diploids are very useful homozygous breeding lines, while the high-yielding triploids and tetraploids have a lot of potential for exploitation in commercial production of seedless fruits (Rimberia et al. 2009).

Polyploidy has received considerable attention in papaya breeding programme. Hofmeyr (1942, 1945) was able to induce polyploidy in papaya. He found that the quality of tetraploid fruit was better than the diploid and the fruit was also compact with smaller seed cavity, but tetraploids were observed to be less fertile than diploids. Singh (1955) reported complete sterility in both female and male tetraploids and expressed doubt about their commercial utilisation. Later, Zerpa (1957) reported colchicines-induced tetraploid hermaphrodite plants which were used as male parent in a cross with a female diploid and the tetraploid produced a few seeds without endosperm. By embryoculture two diploid plants were obtained which turned out to be hermaphrodite.

2.7.6 Mutation Breeding

Bankapur and Habib (1979) attempted mutation in papaya through radiation and observed that doses of gamma rays ranging from 5 to 15 kR were able to produce significant changes in characters such as seed germination, survival of seedlings and ratio of male to female sexes in the population.

Ram (1983) and Ram and Srivastava (1984) exposed dry papaya seeds to 10–70 kR gamma rays and observed that the percentage germination and emergence decreased significantly with increasing doses, and 50 kR and above doses were lethal. Ram and Majumder (1981) evolved a dwarf mutant line by treating papaya seeds with 15 kR gamma rays. Initially, three dwarf plants were isolated from M_2 population. Repeated sib-mating among the dwarf plants helped in establishing a homozygous dwarf line which was later named 'PusaNanha'. It is dwarf (106 cm) in height having a thinner trunk girth (25 cm) and shorter leaf length (86 cm) as compared to 213 cm tall parent plants having thicker trunk (36 cm) and larger (193 cm) leaves. The fruiting in this strain starts at lower height (30 cm). Thus, it is more suitable for high density planting (Ram and Majumder 1988). Work on mutation breeding with gamma rays at Pune, Maharashtra (India) did not produce any fruitful results.

2.8 BIOTECHNOLOGY

Papaya plants can be propagated clonally by tissue culture techniques (Litz 1978; Litz and Conover 1978; Rajeevan and Pandey 1983; Reuveni et al. 1990). In addition to leaf, axillary buds and meristem or shoot-tip culture, seed-coat tissues, embryos and anthers have also been used for raising *in vitro* plants (Yang and Ye 1992; Islam and Joarder 1996; Jordan and Velozo 1997). Good survival, more uniformity in sex and higher yield of tissue-cultured plants has been reported under field conditions (Pandey and Singh 1988). Reports on cell suspension cultures of papaya are few. The cultured conditions allowed initiation of a large number of embryos directly from cell suspensions through adventive somatic embryogenesis and indirectly from callus on axillary buds (Jordan and Velozo 1996). Somatic embryos could be germinated on suitable culture medium until plantlets reach a suitable size for transfer to soil. Somatic embryos have also been raised from the callus of root explants after 3 months of culture (Chen et al. 1987). The explants were obtained from 4 to 6 weeks old seedlings of papaya varieties, Solo and Sunrise. Castillo et al. (1997) could produce somatic embryos of papaya cv., Solo in a liquid production system and encapsulated the embryos in two different encapsulation compounds. The encapsulated embryos showed 77.5% germination. The embryogenic regeneration, though obtained efficiently from several kinds of immature tissues in the presence of auxins, is nonetheless often difficult to achieve on selective media employed in transformation procedures. However, improved methods of preparing embryogenic cultures are likely to overcome this bottleneck (Manshardt and Drew 1998).

Genetic variability with respect to disease resistance in *Carica papaya* L. is low. *In vitro* culture techniques have greatly helped in producing mutants and variants that are being used in breeding programmes in different countries. Cybrids with good attributes can be produced by protoplast fusion with other species of *Carica* (Manshardt 1992). Molecular markers, DNA finger printing and genome mapping are being used to identify the species, their mixtures and their contribution to the progenies in the hybridisation programmes (Manshardt 1992; Sharon et al. 1992; Sondur et al. 1996; Magdalita et al. 1998) and also to variation in the sex forms within the species (Nandi and Mazumdar 1990; Stiles et al. 1993; Manshardt and Drew 1998). A RAPD map of the papaya genome has been generated and used to identify markers for sex determination, flowering height and fruit carpelloidy (Sondur et al. 1996). Somsri et al. (1998) have compared RAPD (Randomly Amplified Polymorphic DNA) and DNA amplification finger printing (DAF) to develop molecular markers for sex prediction in papaya. They observed that DAF produced at least five times more bands than equivalent RAPD reactions permitting more efficient prediction. Preliminary analysis for linkage associations indicated that these markers were closely linked to the sex-determining alleles. Conversion of some DAF markers into more convenient SCAR markers proved difficult since DAF bands were difficult to clone (Somsri et al. 1998).

Phylogenic analysis based on isozyme and DNA polymorphisms indicate that papaya is a distant relative of other members in the genus. Jobin-Decor et al. (1997) have reported that *C. papaya* is about 70% dissimilar to other *Carica* spp. Wild species like, *C. pubesceus* and *C. stipulata* are much closer to each other with similarity of 87% by isozyme analysis and 82% by RAPD analysis (Jobin-Decor et al. 1997). In order to produce inter-specific crosses where the parents are taxonomically distant, barriers that prevent successful hybrid production have been overcome by *in vitro* embryo culture. Yung (1986) and Fitch and Manshardt (1990) succeeded in regenerating papaya plants from immature zygotic embryos through somatic embryogenesis. Magdalita et al. (1996) improved embryo-rescue protocol for younger (90 days old) embryos of *C. papaya* L. (clone 2001), and subsequently utilised the technique for efficient production of inter-specific hybrids of *C. papaya* × *C. cauliflora* from 90 to 120-days-old embryos. The relative ease with which *Carica* inter-specific hybrids can be produced by embryo or ovule rescue (Drew et al. 1998) has virtually directed research activity away from protoplast fusion. However, production of dihaploid inbred lines through anther culture is still under-researched area that merits more attention.

Unfortunately, in most countries, papaya suffers from PRSV, limiting its productivity commercially as well as in the backyard (Gonsalves et al. 2007). The developers of the first transgenic papaya envisaged the GE variety as a promising pro-poor product of biotechnology and were eager to collaborate with researchers from around the developing world. Suitable GE, virus-resistant varieties have now been developed for Brazil, Jamaica, Venezuela, Thailand, China and the Philippines among other countries. Yet, in no place outside Hawaii have growers or consumers reaped the benefits of these plants. Gonsalves and her colleagues (2007) highlighted that this technology is particularly suitable for low-income farmers. With regard to consumer demand, the nutritional value of papaya, while important to Hawaiian consumers, is even more crucial in developing countries where papaya is already popular. GE papaya does not require changes in management practices or large capital investments, it does not alter production costs, and access to intellectual property is already being negotiated in several countries in a philanthropic manner (Gonsalves et al. 2007). Because the effects of PRSV have been as devastating in other countries as they were in Hawaii, there is a clear need for a solution, and a demand by increasingly vocal growers.

Papaya plants suffer from a plethora of viral diseases. Papaya ring spot virus (PRSV) is more devastating and can be transmitted by infected seeds used for propagating the genotype (Verma 1996). Since there are fewer virus particles in apical meristems, the tissue culture technique of meristem culture has been developed as an important and effective method for elimination of viruses. However, all regenerated plants must be checked for presence of virus particles by standard ELISA technique or by cDNA probing. Disease free meristem is, thus, an ideal source material for micropropagation.

In the late 1980s, the University of Hawaii began developing a papaya resistant to papaya ring spot virus. To do this, certain viral genes encoding capsid proteins were transferred to the papaya genome. These viral capsid proteins elicit something similar to an 'immune response' from the papaya plant. These new, genetically modified papaya plants are no longer susceptible to infection, allowing farmers to cultivate the fruit even when the virus is widespread. The first virus-resistant papayas were commercially grown in Hawaii in 1999. Transgenic papayas now cover about one thousand hectares, or three quarters of the total Hawaiian papaya crop. Genetically modified papayas are approved for consumption both in the US and in Canada. Several Asian countries are currently developing transgenic papaya varieties resistant to local viral strains (Ray 2002).

Since the PRSV is a major problem in many papaya growing areas, development of genetically resistant cultivars has been greatly envisaged. Papaya ring spot virus (PRSV) is often a limiting factor in the production of papaya worldwide. In 1992, PRSV was discovered in the district of Puna on Hawaii Island, where 95% of Hawaii's papaya was grown. Within two years, PRSV was widespread and causing severe damage to papaya in that area. Coincidentally, a field trial to test a PRSV-resistant transgenic papaya had started in 1992, and by 1995 the 'Rainbow' and 'SunUp' transgenic cultivars had been developed. These cultivars were commercialised in 1998. 'Rainbow' is now widely planted and has helped to save the papaya industry from devastation by PRSV. Transgenic papaya has also been developed for other countries, such as Thailand, Jamaica, Brazil and Venezuela. Efforts to have these papaya deregulated in these countries are ongoing. Farmers quickly planted the transgenic papaya seeds, which were nearly all 'Rainbow' because the farmers in Pune favoured this transgenic cultivar (Gonsalves 1998). Harvesting of 'Rainbow' started in 1999 and grower, packer and consumer acceptance were widespread. The papaya industry had been spared from disaster. Since 1992, when the virus was discovered in Pune, the yearly amount of fresh papaya sold from Pune had gone from 53 million pounds in 1992 to 26 million pounds in 1998. In 2001, Pune papaya production rebounded to 46 million pounds of fresh market papaya. Another important impact has been the dramatic reduction of PRSV inocula in Pune, because infected fields have been replaced by the resistant transgenic papaya and because many abandoned infected fields have since been destroyed. These conditions, along with judicious isolation and rouging of infected plants, have enabled growers to continue to produce non-transgenic papaya, especially to supply the Japan market, which does not yet allow the importation of transgenic papaya. In January 2003, Canada allowed the importation of transgenic papaya. Yet another benefit is that papaya acreage has expanded on Oahu due to the use of PRSV-resistant transgenic 'Rainbow' (or new hybrids that have been derived from 'Rainbow'). Efforts for the development of transgenic papaya have recently started with Bangladesh, Uganda and Tanzania. Efforts in the first four countries (Brazil, Jamaica, Venezuela and Thailand) have resulted in the development of resistant transgenic papaya that is suitable for their country (Cai et al. 1999; Tennant et al. 2002). In fact, some of the transgenic papaya are well advanced in field trials and are moving through the process of deregulation. How fast the process will move, in the light of the current GMO climate, is not known. However, it is clear that the PRSV-resistant transgenic papaya is a practical solution for controlling PRSV, as has been shown in Hawaii. Conventional interspecific hybridisation is being used to cross papayas with resistant species like, *C. cauliflora* (Chen 1992; Louw 1994; Lehane 1996), cultivars like, Cariflora and F_1 hybrid of the cross, Cariflorax Sunrise Solo (Conover et al. 1986; Escudero et al. 1994). Attempts are also under way to create genetically transformed plants resistant to the PRSV and other viruses.

Mahon et al. (1996) have developed a method for the stable transformation and regeneration of a dioecious papaya cultivar by micro-projectile bombardment. The method was developed after investigating both zygotic and somatic embryos as target tissue and optimisation of a number of parameters using transient expression of the *uidA* reporter gene. Caberera-Ponce et al. (1995) have also produced herbicide resistant transgenic papaya plants using zygotic embryos and embryogenic callus as target cells for particle bombardment. More emphasis has been laid on coat protein-mediated protection through the transfer and expression of the PRSV coat protein (*cp*) gene in papaya (Fitch et al. 1992; Lehane 1996; Fitch et al. 1998; Gonsalves 1998; Yeh et al. 1998). In their studies, Gonsalves et al. (1998) and Cai et al. (1999) used 4-week-old somatic embryos for bombardment with particles containing the non-translatable form of the coat protein (*cp*) gene of PRSV HA5-1 and observed that the nine bombarded plates produced as many as 207 kanamycin-resistant clusters over a period of 7 months. A total of 83 transgenic lines expressing the non-translatable coat protein gene of PRSV were obtained from the somatic embryo clusters that originated from immature zygotic embryos. Twenty-five transgenic lines (out of 83) were resistant to the homologous PRSV isolate from Hawaii and some of these were resistant to PSRV isolates outside of Hawaii, including Australia, Taiwan, Mexico, Jamaica, Bahamas and Brazil. Transgenic plants possessing desirable characters particularly *cp* gene have also been produced through *Agrobacterium tumefaciens* mediated genetic transformation (Yang et al. 1996; McCandless 1997; Yeh et al. 1998). The primary results of the field tests of the *cp* transgenic plants have indicated a great potential for the control of PRSV. Such cultivars are expected to give the papaya production a big fillip. Another area which has drawn attention of the biotechnologists is extension of postharvest life with sense and antisense versions of the ACC synthase and ACC oxidase genes (Neupane et al. 1998). Investigations to characterise polyphenol oxidase and other enzymes obtained from fruits of hermaphrodite and female plants have also been undertaken (Cano et al. 1996; Lin et al. 1998) to examine alterations in fruit ripening and postharvest storability. Today many transgenic papaya plants are under evaluation at various experimental sites.

2.9 GENETIC DIVERSITY

Genetic diversity has played a significant role in the development of cultivars suitable for growing under different agro-ecological regions for fruit as well as papain extraction. The papaya germplasm ranges from very primitive types to major commercial types. There are numerous local mixtures everywhere in all papaya growing regions. Collection, evaluation, documentation, conservation and utilisation of papaya germplasm and thereby, contribution to development of varieties are prerequisite for future advancement (Ram 1992). Because of its complex genetic makeup, there are few true cultivars of papaya, which are as uniform in horticultural characters as the cultivars of other herbaceous crops. When seed results from open pollination, it is impossible, in most cases, to obtain selections which are reasonably uniform in flower type and fruit characteristics. Despite the lack of recognised cultivars, growers can maintain satisfactory strains by controlled pollination of selected plants. Parent plants should be carefully selected for early and heavy fruit production of desirable shape and size. A group of Hawaiian papayas referred to as Solo comes closer to deserving cultivar rank than any other types due to its constancy in character expression to a high degree of natural self-pollination of its bisexual flowers (Medina et al. 2003). This, in addition to continuous selection of pear-shaped fruits produced by bi-sexual plants, has maintained Solo character relatively unchanged over the years. Improved selections, such as Sunrise Solo, have resulted from rigorous breeding work.

2.9.1 Germplasm

The papaya germplasm ranges from very primitive types to major commercial varieties. Wide genetic variability is noticed in papaya growing regions due to seed propagation and cross-pollinated nature of the crop. On account of high degree of variability in economic characters, several selections from the existing populations have been made in different papaya growing areas of the world. In India alone, over 10 such selections are available for commercial cultivation. To widen the genetic base, a large number of superior cultivars have been introduced in India from other countries, namely Solo, Sunrise Solo, Waimanalo, Peradeniya, Thailand, Pink-Flesh, etc. Some hybrids are also available for use in further variety improvement programmes.

2.9.2 Description of Varieties

Despite the great variability in size, quality and other characteristics of papaya, there were few prominent, selected and named cultivars before the introduction into Hawaii of the dioecious, small-fruited papaya from Barbados in 1911. It was named

TABLE 2.2
Improved Papaya Varieties in Different Countries

Country	Variety
South Africa	Hortus Gold, Honey Gold
Sri Lanka	Peradeniya
Hawaii (USA)	Solo, Solo-10, Higgins, Wilder, Waimanalo, Homestead
Trinidad	Santa Cruz Giant (hermaphrodite), Cedros (dioecious), Singapore Pink (hermaphrodite)
Indonesia	Semangka (Red Fleshed Fruits); Thialand
Venezuela	Maradol Roja, Cubana, Paraguanera
Florida (USA)	Betty, Washington, Honeydew
Australia	Improved Peterson, Peterson, Bettina
Surinam	Santo 3, Santo 4, Santo 7

Source: Adapted from Ray, P. K., 2002, *Breeding Tropical and Subtropical Fruits*, Narosa Publishing House, Daryaganj, New Delhi, India.

'Solo' in 1919 and by 1936 was the only commercial papaya in the islands. 'Solo' produces no male plants, only female (with round, shallowly furrowed fruits) and bisexual (with pear-shaped fruits) in equal proportions. 'Kapoho Solo'or 'PunaSolo', 'Dwarf Solo' (a backcross of Florida's 'Betty' and 'Solo') and Waimanalo' ('Solo' line 77) were developed and became popular with growers in USA (Table 2.2).

Higgins (Line 17 A): This was introduced to Hawaiian growers in 1974. It is of high quality, pear shaped, with orange-yellow skin, deep-yellow flesh and averages about 0.5 kg when grown under irrigation. In areas or seasons of low rainfall, the fruit remains undersized.

Wilder (Line 25): This is a cultivar admired for its uniformity of size, firmness, small cavity and now popular for export.

Hortus Gold: This is a South African cultivar, launched in the early 1950s, dioecious, early maturing, round-oval shape, golden-yellow fruits and 0.9–1.36 kg in weight. This possessed higher sugar content and disease resistance was chosen and named 'Honey Gold' in 1976. This cultivar had a slight beak at the apex, golden-yellow skin, sweet flavour and good texture, but becomes mushy when overripe. It averages 1.0 kg per fruit, except for those at the end of the season which are much smaller. It does not reproduce true from seed and is, therefore, propagated by cuttings. It is late-maturing (10 months from fruit set to maturity) and, therefore, brings nearly double the price of other cultivars.

Bettina and Petersen: These long-standing cultivars of Queensland, Australia were inbred for several generations to obtain pure lines. 'Bettina', a hybrid from a cross between Florida's 'Betty' and a Queensland strain, is a low, shrubby, dioecious plant producing well-coloured, round-oval fruits weighing 1.5–2.5 kg.

Improved Petersen: This is dioecious, tall-growing variety, with fruits noted for the fine colour and flavour of the flesh. In 1947,'Bettina 100A' was crossed with 'Petersen 170' to produce the superior, semi-dwarf' Hybrid No. 5′, with fruits which were smooth, yellow, rounded-oval, 1.5 kg in weight, thick-fleshed, of excellent flavour and prized for marketing fresh and for canning. It produced better fruits than either of its parents, and remained a preferred cultivar for more than 20 years. 'Solo' and 'Hortus Gold' are often grown but most plantations are open-pollinated mixtures.

Cedro: This is dioecious, rarely bisexual, a heavy bearer and highly resistant to anthracnose. The fruits weigh from 1.5–3.5 kg averaging 2.5 kg; have firm, yellow, melon-like flesh and are suitable for fresh sale or for processing.

Singapore Pink: In this variety, the plants are mainly bisexual, producing cylindrical fruits. The minority are female with round fruit. Average weight of fruits is 2.27 kg though they vary from 1.0 to 3.0 kg. The flesh is pink. The fruit surface is prone to anthracnose in rainy season, so the fruits, at such times, must be picked and sold in the green state. Two smaller-fruited types, 1.0–1.5 kg in weight, with bright-yellow skin and thick, firm flesh were selected for marketing fresh.

Cariflora: This is a new cultivar developed at the University of Florida, Homestead. It is nearly round, about the size of a cantaloupe (musk melon), with thick, dark-yellow to light-orange flesh; tolerant to papaya ringspot virus, but not resistant to papaya mosaic virus or papaya apical necrosis virus. Yield is good in southern Florida and warm lowlands of tropical America but not at elevations above 2625 ft (800 m).

Sunrise Solo: This was formerly known as HAES 63-22 and introduced from Hawaii into Puerto Rico. The fruit has pink flesh with high total solid content. In Puerto Rican trials, seeds were planted in mid-November, seedlings were transplanted in the field 2 months later, flowering occurred in April and mature fruits were harvested from early August to January. The selections from Puerto Rican breeding programmes are 'P.R. 6-65' (early), 'P.R. 7-65' (late), and 'P.R. 8-65'.

UH Rainbow: This is a new hybrid developed at the University of Hawaii at Manoa. 'UH Rainbow' combines the superior quality typical of Hawaii's 'Solo' papayas with excellent resistance to a devastating plant virus disease-papaya ringspot virus (PRSV). This combination of traits was accomplished through genetic engineering, one of the latest advances in agricultural biotechnology. 'UH Rainbow' is F_1 (first-generation) hybrid produced by crossing Hawaii's standard export variety, 'Kapoho' with the first genetically engineered papaya possessing resistance to PRSV, 'UH SunUp'. The resulting hybrid is an excellent source of vitamins A and C and is highly productive. The fruits are pear shaped (with short neck) to elliptical, 650 g fruit weight, yellow orange flesh colour, TSS 12-16° Brix, uniform and juicy texture and mild aroma. The fruit is ready to eat when it is 70%–90% yellow.

In India, papaya breeding and selection work has been carried on for over 30 years beginning with 100 introduced strains and 16 local variations (Table 2.3). A well-known cultivar is 'Coorg Honeydew', a selection from 'Honeydew' at Chethalli Station of the Indian Institute of Horticultural Research (IIHR), Bangalore (Karnataka). There are no male plants; female and bisexual plants occur in equal proportions. The plant is low-bearing and prolific. The fruit is long to oval, weighs ranging from 2.0 to 3.5 kg and possesses yellow flesh with a large cavity.

'Washington', popular in Bombay, has dark-red petioles and yellow flowers. The fruits are of medium size with excellent, sweet flavour. 'Burliar Long' is a prolific variety, bearing as many as 103 fruits in the first year, mostly in pairs densely packed

TABLE 2.3
Important Papaya Varieties Cultivated in Different States of India

State	Varieties
Andhra Pradesh	Honeydew, Coorg Honeydew, Washington, Solo, CO-1,CO-2, CO-3, Sunrise Solo, Taiwan
Jharkhand and Bihar	Ranchi selection, Honeydew, Pusa Delicious, Pusa Dwarf and Pusa Nanha
Karnataka and Kerala	Coorg Honeydew, Coorg Green, Pusa Delicious and Pusa Nanha
West Bengal	Ranchi selection, Honeydew, Washington, Coorg Green
Orissa	Pusa Delicious, Pusa Nanha, Ranchi selection, Honeydew, Washington, Coorg Green

Source: Adapted from Ray, P.K., 2002, *Breeding Tropical and Subtropical Fruits*, Narosa Publishing House, Daryaganj, New Delhi, India.

FIGURE 2.6 **(See colour insert.)** Important papaya varieties from IARI: (a) Pusa Majesty, (b) Pusa Delicious, (c) Pusa Giant, (d) Pusa Dwarf and (e) Pusa Nanha.

along the stem down to 45 cm from the ground. Seedlings are 70% females and bloom 3 months after transplanting.

A papaya breeding programme of the Indian Agricultural Research Institute at Regional Station, Pusa (Bihar), initiated in 1966, has led to the release of five varieties (Figure 2.6):

Pusa Majesty: A gynodioecious line is tolerant to viral diseases and root knot nematodes (Ram 1984a). The variety is suitable for papain production and is comparable to CO-2 variety for papain yield. The fruits are medium sized, 1.0–1.5 kg in weight, round in shape and have better keeping quality. It starts fruiting after 146 days from the time of transplanting.

Pusa Delicious: It is a selection from progenies of variety 'Ranchi', gynodioecious, excellent fruit quality, deep orange flesh; 13% TSS with medium-tall plants and starts yielding 8 months after planting. The fruit is medium sized (1.0–2.0 kg) with excellent flavour. It is grown as a table purpose variety (Ram 1984a).

Pusa Nanha: A dioecious is dwarf mutant and precocious variety of papaya which is suitable for pot cultivation. Fruiting starts at a height of 40 cm within 239 days of planting with the total height of the plant being 130 cm. Fruits are medium to small, oval and flesh 3.5 cm thick of blood red to orange colour with TSS ranging between 6.5 and 8.0° Brix.

Pusa Dwarf: It is a dioecious variety with dwarf plants and medium-sized (1.0–2.0 kg) oval fruits. The plant starts bearing from 25 to 30 cm above the ground and is comparatively drought hardy. This variety is highly suitable for high-density planting (Ram 1981).

Pusa Giant: This is a dioecious line, most vigorous and produces large-sized fruits (2.5–3.0 kg each). It starts fruiting at a height of 92 cm within 259 days of planting. The plant reaches a height up to 220 cm. It is sturdy and tolerant to strong winds. Fruits are oblong with 18 × 10 cm seed cavity and 1.5 cm thick flesh of orange colour having TSS between 7 and 8.5° Brix. It is suitable for marketing ripe or green for use as a vegetable and also for canning (Ram 1984a).

Tamil Nadu Agricultural University (TNAU), Coimbatore has also started work on papaya improvement at Coimbatore. It has released seven varieties so far which are discussed in the following (Figure 2.7):

CO-1: It is a selection from progenies of var. Ranchidone at TNAU, Coimbatore in 1972. The plant is dwarf in height, producing the first fruit within 60–75 cm from the ground level. Fruit is medium sized, spherical, round or oval with smooth greenish-yellow skin and orange-yellow, soft and firm flesh. It is moderately juicy with good keeping quality. The objectionable papain odour is practically absent in the fruits.

CO-2: It is a pure-line selection from a local type at Agricultural College and Research Institute, Coimbatore done in 1974. Plant is medium tall; fruits are medium sized, obovate, greenish yellow and ridged at the apex and flesh is red, soft to firm and moderately juicy with good keeping-quality. It is a suitable type for extraction of papain. It gives 4–6 g dried papain/fruit or 250–300 kg papain/ha.

CO-3: This is hermaphrodite (55 female: 45 hermaphrodite flowers) type hybrid developed during 1983. The fruit of this hybrid (CO-2 × Sunrise Solo) is larger in size when compared with Solo and exhibits all the desirable attributes of Solo. Total soluble solids (TSS) are as high as 13.8° Brix and average fruit weight ranges from 1.0 to 1.5 kg with reddish flesh colour. The fruits have a good keeping quality. Each tree yields 100–120 fruits in 2 years.

CO-4: This is a hybrid between 'CO-1' × 'Washington' released in 1983. Plant parts have purple colouration. The fruit is medium in size and round in shape similar to 'CO-3'. Flesh colour is with purple tinge. About 80–90 fruits/plant are obtained during the bearing period of 2 years. The fruit quality is better than 'CO-1' and 'Washington'.

CO-5: It is a selection made in 1985 from Washington and isolated for its high papain production. It produces consistently 14–15 g dry papain/fruit. It gives 75–80 fruits/tree in 2 years with an average yield of 1500–1600 kg dried papain/ha.

FIGURE 2.7 (**See colour insert.**) Important papaya varieties from Tamil Nadu Agricultural University, Coimbatore: (a) CO-1, (b) CO-2, (c) CO-3, (d) CO-4, (e) CO-5, (f) CO-6 and (g) CO-7.

CO-6: This is a dioecious variety with dwarf stature, large fruit size, good for papain production and is a selection from Giant made in 1986. The fruits are larger in size weighing 1.5–2.0 kg. Fruits are round to oval in shape with yellow flesh. Tree yields 80–100 fruits. It is suitable both for table purpose and latex extraction.

CO-7: A hybrid between Coorg Honeydew × CP85; high yielding, red fleshed, uniform in fruit shape, gynodioecious with good edible character, least summer skip and stamen carpellody, yields 98 fruits/tree each weighing 1.15 kg, and was released in 1997. This variety is similar to CO-3 but the fruit size is bigger. It yields 65–70 fruits per tree. It is obtained from a multiple cross involving, Pusa Delicious, CO-3 and Coorg Honeydew as parents.

A systematic papaya breeding programme was started in 1972 at G.B. Pant University of Agriculture and Technology, Pantnagar, India leading to the development of three superior varieties:

Pant Papaya-1: It is a vigorous dwarf selection bearing first flower at a height of 45–60 cm with a total height of 125–135 cm in the first year.

Pant Papaya-2: Plants are vigorous with medium height (150–220 cm) in the first year bearing, first flowering starts at a height of 90–100 cm from the ground.

It produces medium to large size fruits (1.0–2.0 kg each) of good quality with high yield potential of 30–35 fruits per plant. It is tolerant to both frost and water logging.

Pant Papaya-3: The plants are of medium height (225–250 cm) in the first year, bearing first flower at a height of 115–130 cm from the ground on strong stem. Fruits are of small to medium size (0.5–0.9 kg each) with excellent quality. It bears large number of fruits per plant (45–60) although of small size and the plants are tolerant to frost and water logging. Due to small size of fruits, it can be grown for distant and other export markets.

Washington: Trees are fairly vigorous, tall stem with purple rings, dark purple petiole growing darker towards the lamina, yellow deep flowers, medium to large size fruits (1.0–1.5 kg each) of ovate to oblong shape having distinct purple colour ring at its top connected with the fruit stalk. Fruit pulp is yellowish red, very sweet with agreeable flavour of fine consistency, few seeds and has better keeping quality. The flesh is free from the disagreeable papain, making it most popular variety.

Coorg Honeydew: Popularly known as 'Madhubindu', it is cultivated for table as well as processing purpose. Due to its excellent fruit quality, it fetches good market value. As a gynodioecious type, it produces both andromonoecious (male and bisexual flowers in a tree) and pistillate plants. It is a selection from Honeydew bred by ear-to-row method of selection and developed at Chetalli Research Station, Indian Institute of Horticultural Research, Bangalore, Karnataka. Since every plant bears fruits, planting of more than one seedling per pit is not required. Its trees are of medium height and bear fruit heavily low on the trunk in the first year. The fruits are medium to large (1.25–2.5 kg), oblong to long, smooth skinned, slightly beaked and prominently ridged at the apex, orange fleshed, soft, moderately juicy and have good taste. The fruits have big central cavity and are of moderate keeping quality.

HPSC-3: This is a selection from the cross Tripura Local × Honeydew. This hybrid produces higher yield (197.7 t/ha) compared to both the parents, performs well for papain production (5 g/fruit)and shows resistance to papaya mosaic.

Surya: This is a selection from the cross, Sunrise Solo × Pink Flesh Sweet developed at the Indian Institute of Horticulture Research (IIHR), Bangalore (Karnataka). It is a gynodioecious hybrid. Plants are dwarfer than Sunrise Solo. Fruits are medium sized, weighing on an average 600–800 geach. The pulp is red with a thickness of about 3.0–3.5 cm. Fruits are sweet in taste with a TSS of 13.5–15.0° Brix.

Ranchi: It is a variety from Jharkhand and popular in south India. The fruits are oblong with dark yellow pulp and sweet taste.

IIHR 39 *and IIHR* 54: These varieties were developed at IIHR, Bangalore and bear medium-sized sweet fruits with high TSS (14.5° Brix) with better shelf life.

Pink-flesh Sweet: Medium-sized pink-fleshed fruits of this variety possess TSS ranging from 12° to 14° Brix.

Punjab Sweet: It is a selection made at PAU, Ludhiana. The plants are dioecious and frost resistant. Fruits are large and oblong, weighing more than 1.0 kg each and yield is 50 kg/plant.

Ram and Majumder (1984) and Ram et al. (1985a) collected more than 125 genotypes of papaya and *Carica* species at Pusa, Bihar, from 1966 to 1992.

The component characters, namely earliness, fruit yield, fruiting height, single fruit weight and the height of plant were found to be important for the expression of genetic diversity and in the selection of parents for improvement programme (Ram and Majumder 1992).

Varieties/hybrids developed in the private sector

Taiwan-785: This variety is cultivated for table as well as processing purposes. The plant is dwarf in stature, producing the first fruit within 60–75 cm from the ground level. Fruits are oblong with thick, orange-red and sweet pulp. Each tree yields 100–125 fruits in 1 year. It has a good keeping quality and disease tolerance.

Taiwan-786: It is a gynodioecious variety cultivated for table as well as processing purposes. The fruits are oblong with a tasty sweet pulp having few seeds. The plant starts bearing fruits from 100 cm above the ground level. The fruit weighs between 1.0 and 3.0 kg and has excellent keeping quality.

F_1 *Red Winner*: It is a high-yielding orange to red fleshed papaya hybrid. All plants bear fruits. Female fruits are slightly rounded and hermaphrodite fruits are oblong. Average fruit weight is approximately 1.5–2 kg. The firm flesh is orange to red, with very strong flavour and has a high sugar content of 14%. This variety is fairly tolerant to papaya ring spot virus (PRSV).

F_1 *Aroma*: This is a popular Hawaiian type papaya hybrid. Fruits are oval pear-shaped weighing around 0.7–0.8 kg each. The papaya flesh is orange to red, has a high sugar content and strong flavour. The fruits are suitable for export markets.

F_1 *Taiwan Princess*: This is a very popular Taiwan type papaya hybrid, widely grown in USA, Mexico, Argentina, Peru, Taiwan, China, India and Australia for export and local markets. Fruits are large with pointed blossom end and weight around 1.2 kg each. Flesh is orange red and tender with good taste and quality. The variety is tolerant to papaya ring spot virus (PRSV).

F_1 *Carina*: Carina is a dwarf high yielding variety widely adapted to various agro climatic conditions. The proportion of hermaphrodites to female plants is 1:1. The flesh is thick, reaching a yellow-orange skin colour when the fruit becomes ripe. The pulp colour is red salmon, solid and sweet with 11–12° Brix. The fruit weighs around 1.6–2.5 kg. The fruits are highly prized because of its consistency and flavour. Under ideal conditions, F_1 Carina can yield around 225–250 tons/ha.

F_1 *Dawn*: Plants are dwarf making it easier for spraying and harvesting. F_1 Dawn has a production potential up to 300 tons/ha. The proportion of hermaphrodites and female plants is 1:1. The flowering period ranges from 42 to 46 days after transplanting it. On an average, each plant bears around 55–60 fruits with an average weight of 3.3 kg. The flesh is reddish-orange colour, with a thickness of 3.25 cm and a sweetness of 12° Brix.

F_1 *Herma*: This is a prolific and vigorous hybrid with red-fleshed fruits weighing 1.1–1.25 kg. Fruits set at 75 cm (30 in) from ground. It is good shipper with hard outer shell, sweet flavoured and suitable for both table (home consumption) and processing purposes. It produces 1:1 ratio of hermaphrodite to female plants. Female fruits are globe shaped and hermaphrodite fruits have a pointed blossom end. The variety is tolerant to papaya ring spot virus (PSRV).

F_1 Yellow Fellow: This is a prolific heavy yielder with thick yellow fleshed fruits weighing 2.5–3.5 kg. Fruits set about 80 cm from ground level and flowers at 45–50 days after transplanting. The hybrid is yellow skinned, suitable for both table (home consumption) and processing purposes for the papain industry. It produces 1:1 ratio of hermaphrodite to female plants with all plants bearing fruits of sweet flavour. Its TSS content is 12° Brix.

3 Climate and Soil

3.1 TEMPERATURE

Papaya is a tropical plant and very sensitive to frost, limited to the region between 32° north and 32° south of the Equator. It is grown up to an elevation of 1200 m above the mean sea level. It needs plentiful rainfall or irrigation, but must have good drainage. Flooding for 48 h is fatal to it. Brief exposure to 32°F is damaging; prolonged cold without overhead sprinkling can kill the plant (DAIS 2009). Although a mature papaya tree can withstand a temperature of –2°C, production is only recommended in areas where the average daily minimum temperature during mid-winter never drops below 5°C. Ideally, night temperatures should not drop below 12°C. The optimum temperature range for papayas is between 25°C and 30°C. Temperatures higher than 36°C and lower than 16°C for extended periods can adversely affect the growth of the trees. It grows well in regions where summer temperature ranges from 35°C to 38°C. It requires 23°C average daily temperature and 6 months for flowering to fruit maturity. Although papayas are considered sun-loving plants, morphological plasticity in the shade is high and involves changes in many characteristics such as leaf mass per area, chlorophyll *a* and *b* ratio, stomata density, internode length and degree of blade lobbing (Buisson and Lee 1993). This plasticity is evidenced by the morphology adopted by papayas growing in multi-storied agro-ecosystems and in high-density orchards as well (Iyer and Kurian 2006). The lower the temperature, the longer the time it will take for fruit maturity with round fruit shape. After the winter, trees are in a recovering stage and produce few flowers. In subtropical conditions, however, growth and flower production cease when the night temperature drops below 12°C.

Papaya plants exhibit C3 type photosynthesis systems (Campostrini and Glenn 2007). Optimum temperature for growth is 21–33°C, under which papayas can produce two leaves per week and 8–16 fruits per month. Temperatures below 10°C are not well-tolerated (Allan 2005). Light compensation point for leaf-level photosynthesis is Ca. 35 μmol m^{-2} s^{-1}, and saturation is reached at Ca. 2000 μmol m^{-2} s^{-1} of photosynthetic photon flux density (PPFD) (Campostrini and Glenn 2007). High photosynthetic rates of 25–30 μmol CO_2 m^{-2} s^{-1} alternate with pronounced (midday) gas exchange depressions, apparently caused by direct stomatal and cuticular responses to air humidity (Marler and Mickelbart 1998). This process may reduce productivity by 35%–50% (Campostrini and Glenn 2007).

3.2 WIND VELOCITY

Papaya does not like strong, cool, hot and dry winds. In open and high-lying areas, plants are exposed to strong winds or storm. Papaya seedlings and adults are very responsive to mechanical stimuli and show strong thigmomorphic responses or touch-regulated phenotypes (Porter et al. 2009). These responses could be essential to the success of papaya in harsh, early successional sites exposed to high winds because it triggers hardening mechanisms that result in compact architecture, increased lignification and the formation of petiole cork outgrowths (Clemente and Marler 2001). Therefore, for proper establishment of papaya plantation, it is better to grow under shelterbelt with full sunshine. Staking, earthing up and wind break can decrease the damage to plants from strong winds.

3.3 SOIL

Papayas grow and produce well on a wide variety of soils. The trees often develop a strong taproot shortly after planting. Under favourable conditions the root system can penetrate soil up to a depth of 2 m, but most of the roots responsible for nutrient uptake are found in the top 50 cm of soil, with higher concentrations in the top 25 cm soils. Well-drained sandy loam soil with adequate organic matter is most important for papaya cultivation. On rich organic soils, papaya makes lush growth and bears heavily, but fruits are of low quality (DAIS 2009). Well-drained soils of uniform texture are highly preferable to avoid collar rot disease too. In a high rainfall area, if drainage is poor and roots are continuously drenched for 24–48 h, it may lead to the death of the plants. Sticky and heavy soils are not suitable as rainwater may accumulate in the soil even if it rains only for a few hours. In such cases, higher raised bed and drainage ditch are recommended (Figure 3.1).

A well-drained/upland field is selected for cultivation, as shown in Figure 3.1. The growing field should be well levelled and kept at a suitable soil moisture that is

FIGURE 3.1 (See colour insert.) Raised beds for papaya plantation.

necessary for the growth of papaya plants, although dry climate at the time of ripening is good for the fruit quality. Continuous cropping in the same field may result in poor growth and cause disease problems for papaya trees. The ideal soil texture for papaya cultivation under irrigation is a sandy loam or loam soil, but soils with a clay content of up to 50% are also suitable. In very sandy soils, temporary over-saturated conditions may occur if soil compaction or impermeable layers limit drainage. Sandy soils normally have a very low water-holding capacity and poor nutrient status due to low organic matter and clay content (<10%). A mulch and organic material application can increase the potential of such type of soils. The ideal soil has a fairly loose, brittle and crumbly structure. A compact or strongly developed soil structure adversely affects water infiltration and root penetration. These soils are normally associated with very high clay content in the sub-soil (>50%). While doing best in light, porous soils rich in organic matter, the plant will grow in scarified limestone or various other soils only with adequate care. It is preferred to plant a cover crop (legume) as a source of organic matter. In such cases, the crop must be planted about 6 months before the transplanting (DAIS 2009). Calcium and phosphorous are elements that move very slowly downwards in soils. If there is a shortage of one of these elements, especially in the sub-soil, it should be incorporated during soil preparation. If lime needs to be applied, it should be incorporated into the soil 6 months to a year before planting. Papayas grow best in soils with pH values of 6.0–6.5. If the soil exchangeable aluminium (Al) is not more than 30 ppm, soils with a pH of 5.5 or higher may be used (Rex and Rivera 2005). Decreased yields of papaya due to nutritional deficiency of nitrogen, phosphorus, potassium, calcium and magnesium in acidic soils (pH values below 5.5) have been reported. To overcome this problem, either calcium carbonate or lime is added to the soil. On the other hand, if pH of the soil is above 8.0 (alkali), deficiencies of magnesium, manganese, iron, zinc, copper and boron have been reported. In this case, addition of sulphur is very effective (Medina et al. 2003).

4 Cultural Practices

4.1 INTERCULTURAL OPERATIONS

Papaya orchards should be free from unwanted plants or weeds for better growth and development. Weeds must be controlled, especially during the initial stages of orchard establishment. Use of herbicide, hand weeding, mulching and deep hoeing, either singly or in combination is recommended during the first year to check weed growth. Weeding should be done on regular basis, especially around the plants. Applications of Fluchloralin or Alachlor or Butachlor (2.0 g/ha) 2 months after transplanting can effectively control the weeds for a period of 4 months. Mixture of oryzalin + oxyfluorfen is a broad spectrum herbicide which is used in papaya fields for weed management, but the use of oryzalin is safe around young plants but should not be applied to green stems. Oxyfluorfen is also effective as a pre-emergence with sufficient irrigation, but sometimes causes vapour drift to recently emerged leaves. Use of cover crops can also be a practical method to control weeds. Mechanical cultivation between rows is appropriate to disturb the shallow roots. The organic material such as crop residues, farm yard manure, by-product of the timber industry may be used with 5–15 cm layer and keep mulch 30 cm away from the trunk. Inorganic materials such as plastic films of different colours and thicknesses are used efficiently. Mulching in papaya helps by reducing soil compaction, soil erosion, leaching of fertilisers, winter injury, weed problems, retain soil moisture and improve the quality of produce. Papaya plants are dependent on mycorrhizas for their nutrition and benefit from soil mulching and appropriate drainage that facilitate biotic interactions in the rhizosphere and water and nutrient uptake, especially phosphorus and nitrogen (Jimenez et al. 2014). Four genera and 11 species of arbuscular mycorrhizal fungi have been reported associated with papaya roots: *Glomus*, *Acaulospora* and *Gigaspora*, among others (Khade et al. 2010). Mycorrhizal interactions of male and female papaya plants may differ: females seem more responsive to changes in soil fertility and readily adjust mycorrhizal colonisation accordingly (Vega-Frutis and Guevara 2009). Among different mulching material and methods, namely organic mulch, plastic mulch, bare soil with mounding and flat ground methods, the best treatment combination was plastic mulch with mounding, which significantly differs from the remaining for management of phytophthora (root rot) disease.

4.2 EARTHING UP

Earthing up is done three times, for example, after 4th, 6th and 8th months of transplanting, before the onset of monsoon and after ensuring one plant per pit, to avoid overcrowding or water logging. Earthing up in 30 cm radius around the plants also helps the plants to stand erect in sandy loam soils. It is also carried out just after top dressing in the plants on or before the onset of monsoon to prevent the fertilisers losses through volatilisation and seepage (Saran et al. 2014a).

4.3 STAKING

Staking is done with bamboo and other sticks as per availability during fruit development stage to protect the plant from stormy winds and handle fruit load especially after heavy rainfall and irrigation (Figure 4.1). Plants are tied with rope, especially when bearing heavy fruits, and also to support the leaning trees to keep them straight and upright. If earthing up is done properly, the need to stake the plants can be done away with. Planting of shelterbelts of banana, dhaicha (*Sesbania*), and so on, can be done on the windward side of the orchard in order to prevent from strong winds (Saran et al. 2014a).

4.4 FRUIT THINNING

Thinning of fruits becomes essential in papaya when several fruits set in a cluster instead of a single fruit at each node. Some dwarf cultivars like Pusa Dwarf and Pusa Nanha had a shorter internodal space. The operation is performed immediately after fruit set and involves thinning or removal of the fruit clusters, leaving only one fruit at each node. Thinning the fruits that are poorly pollinated, malformed and infected, prevents the damage of fruits due to over-crowding, avoids pressure injury and competition amongst them for space and nutrients (Figure 4.2). Thining of small fruits also should be done, if the trees are severely damaged (Saran et al. 2014a).

FIGURE 4.1 (**See colour insert.**) Heavy-bearing papaya needs staking to prevent lodging.

FIGURE 4.2 (**See colour insert.**) Papaya fruit size: (a) after thinning and (b) without thinning.

4.5 ROGUING

As soon as the plants start flowering, the extra male plants are uprooted (Figure 4.3). Weaker and diseased plants should be uprooted, after ensuring one plant/pit in dioecious varieties. About 5%–10% of the male plants are kept in the orchards for adequate pollination. The hermaphrodite plants produce complete flowers and should not be confused with male while removing them from the orchard (Saran et al. 2014a).

4.6 FRUIT COVERING

At a very early stage, papaya fruits are infested with papaya fruit fly. Bagging with cheesecloth, glassine paper or wax paper bags can be an effective control measure for the fruit fly in small plantings. Bagging should begin when the fruit is small; shortly after the flower petals have fallen off (crown stage). Each fruit should be

FIGURE 4.3 (a) Three plants per pit in dioecious and (b) roguing at flowering stage in papaya.

enclosed in a paper bag or rolled tube of newspaper and tied around the stem. This method can be very practical and successful if enough labour is available. Growing fruits are also covered with paper to avoid the sun scald.

4.7 NIPPING

Cold weather may interfere with pollination and cause shedding of unfertilised female flowers. After harvesting first crop, the terminal growth may be nipped off to induce branching, which tends to dwarf the plant and facilitates easy harvesting of succeeding crops. Removing the side shoots of the stem during the first crop and cutting the old, dry, diseased leaves and petioles are regular practice for a healthy harvest. Transmission of the virus mechanically from infected plant to others should be avoided while carrying out these practices.

4.8 POLLARDING

Mainly, congenial conditions for papaya growth in subtropical parts make plants tall and thin, and they often get broken during cyclones and heavy rains. The fruit yield is also poor which makes orchard uneconomical after the first-year crop of the papaya. Hence, pollarding is required for ratoon crop (1-year-old) in order to reduce the plant height and enhance yield and productivity. Prakash et al. (2014) reported that the pollarding operation is done in the month of January at a height of 60 cm under field conditions due to early sprouting, flowering, fruiting and highest yield (34.56 kg/plant), while number of sprouted buds was maximum under 90 cm pollarded plants and it was minimum under 15 cm. The survival percentage was very poor under lower height of the pollarding. This indicates that the sprouting buds are active in the middle and upper zones of 1-year old papaya plants. Both higher and lower pollarding heights did not produce encouraging results due to shattering of branches at fruiting stage (Figure 4.4). The retention of two sprouted buds was ideal in terms of fruit yield. The retention of higher number of sprouted branches was recommended for pruning during fruit growth and development.

4.9 SMUDGING

Cold weather may interfere with pollination and cause shedding of unfertilised female flowers. Spraying the inflorescence with growth regulators stops flower drop and enhances fruit set significantly. After the first crop, the terminal growth may be nipped off to induce branching which tends to dwarf the plant and facilitates harvesting. However, unless the plants are strong enough, fruiting branches may need to be propped to avoid collapse.

4.10 CROPPING SYSTEMS

The papaya-based agro-horticultural systems are self-sustainable systems where solar energy can be harvested properly and soil resources can be efficiently used.

FIGURE 4.4 (**See colour insert.**) Position of sprouting buds in one-year-old pollard papaya plant.

The systems consist of two main components, namely main crop and intercrop. Papaya is efficient enough in providing higher economic returns even under unfavourable conditions prevailing under the upland situations than other annual crops. The approach aims at improving productivity by the effective utilisation of air space that is not utilised in a single tier system. The multitier system aims at sustainable management of natural resources like soil, water, space and environment. The spatially differential root distribution of different component crops in the system helps in higher nutrient and water use efficiency of the multitier system allowing feeder roots of the component crops at different depths. Again, the increase in organic carbon content by the soil due to decomposition of fallen leaves from the fruit trees contribute towards enhanced biological activity in the system leading to environmental stability in the rhizosphere (Solanki et al. 2013).

The crop rotations of leguminous crops after non-leguminous ones and shallow rooted crops after deep-rooted ones are beneficial. No intercrops are taken after the onset of flowering stage in papaya without efficient INM (Integrated Nutrient Management). In the beginning, sufficient space is available in the orchard, and, therefore, some crops can be taken up with the advantage. Resource degradation leading to an unsustainable production system has demanded our attention for sustainable practices to assure continued production. Intercropping of elephant foot yam in papaya like in banana, coconut and other newly planted orchards gives additional income to farmers. Traditionally, intercropping in interspaces of fruit orchards is practiced due to economic consideration.

In this context, papaya-based agri-horticultural system has immense potential for the betterment of poor farmers. Papaya orchards are planted at 1.8 × 1.8 m spacing

TABLE 4.1
Income from Papaya Based Intercropping Systems

Intercropping System	Yield (q/ha)	Total Income (Rs./ha)	Production Cost (Rs./ha)	Net Return (Rs./ha)
Papaya sole crop	917.00	275,100	100,000	175,100
Papaya seed production	0.40	1,600,000	128,000	1,472,000
Papaya + Turnip + *Suran*	900 + 25 + 195	392,500	172,000	220,000
Papaya + Lentil + *Suran*	900 + 8 + 195	383,500	170,000	213,500
Papaya + Turmeric	900 + 310	401,750	186,000	215,750

1. Rate of turnip = Rs. 10/kg.
2. Rate of papaya fruit = Rs. 3/kg.
3. Rate of lentil = Rs. 20/kg.
4. Rate of turmeric = Rs. 4.25/kg.
5. Yield of fruit/plant = 33 kg.
6. Number of papaya plants/ha = 2778.
7. Rate of papaya seed = Rs. 40,000/kg.

and the inter-spaces may be utilised by growing profitable intercrops. Generally, the inter-crops grown are turmeric, ginger, lentils, turnip, tobacco and so on. The overall production significantly increases due to intercrops, and it was observed to be maximum in papaya in association with turnip and *suran* (elephant foot yam), followed by turmeric and *suran*. Economic analysis of the systems in terms of benefit: cost ratio revealed that 'papaya + turmeric' gives a higher income, followed by 'papaya + ginger' and 'papaya + turnip' (Table 4.1). The interspaces of the papaya orchard could be utilised for growing various intercrops to generate substantial additional income without adverse effect on the soil fertility and productivity of the main crop (Saran et al. 2013).

Among the intercrops, turmeric, turnip and *suran* were found to be the most profitable root crops under partial shade of papaya orchards. The papaya–*suran* intercropping system needs more space, and it may be grown only in autumn transplanted crop of papaya. Under autumn planting of papaya crop, turnip is the best option for NEPZ (North Eastern Plains Zone of India) farmers during the initial period of 3–5 months (Figure 4.5a). Lentil is also a good option in papaya orchard during *rabi* season (Figure 4.5b). *Suran* is a long duration crop (Figure 4.5c). Under irrigated conditions, papaya orchardists may grow *suran* in summer (March) which attains maturity by November.

Papaya-based cropping systems (sequential and intercropping) are found to be the most remunerative as they give high net returns in case of papaya + tobacco intercropping in North Bihar (Ram and Pandey 1990). Net return to the tune of Rs. 1, 67,000 per hectare in 1 year 8 months in case of papaya + tobacco intercropping was also reported (Table 4.2).

FIGURE 4.5 (**See colour insert.**) Papaya on raised bed with intercrops: (a) turnip, (b) lentil and (c) suran.

4.10.1 Advantage of Intercrops

Papaya orchard should be free from unwanted plants for better growth and development. Weeds must be controlled, especially during the initial stages of establishment. Herbicides, hand-weeding, mulching and deep hoeing are usually recommended to check weed growth during the first year, but they are costly. Uses of intercrops like lentil, turmeric and *suran* to cover inter-row spaces are some of the practical methods that reduce the need for labour or herbicide spray for control of the weeds.

TABLE 4.2
Income from Papaya-cum-Tobacco Intercropping

Intercropping System	Yield (q/ha)	Total Income (Rs./ha)	Production Cost (Rs./ha)	Net Return (Rs./ha)	Total Net Income (Rs./ha)
Tobacco sole crop	20	60,000	14,524	45,476	45,476
Papaya sole crop	900	135,000	35,000	100,000	100,000
Tobacco + Papaya	20 + 900	60,000 + 135,000	14,524 + 13,476	45,476 + 121,624	167,000

1. Rate of tobacco (Dried leaves) = Rs. 30.00/kg.
2. Rate of papaya fruit = Rs. 1.50/kg.
3. Fruiting plants in 1 ha = 2,250.
4. Yield of fruit/plant = 40 kg.

4.11 COMPETITION WITH TRADITIONAL CROPS

Papaya is in huge competition with traditional fruit crops such as banana, pineapple, grapes and guava. Despite its nutritional and medicinal importance, papaya finds restricted place in the farming systems and is not being exploited commercially. The traditional fruit crops have established their place in farming systems, developed

FIGURE 4.6 (**See colour insert.**) Collar rot disease–free papaya on raised bed.

fairly good liking and market and extension support, whereas papaya is confined to home yards or near irrigation sources of farming communities. Hardly 2–5 plants are seen at these places, and it is a usual practice with most of the farmers. The common uses for which papaya is grown are raw fruit as vegetable and ripe fruit as medicine for curing stomach disorder. The reasons for its exploitation for gainful employment from production to processing. The major constraints for its commercialisation are lack of proper market, farmers' perception of risk, complex crop, lack of technology, lack of extension support, misconceptions about or prejudices against papaya within society and so on. The research and extension efforts are required to make this crop competitive and relatively advantageous for generating income and employment.

4.12 TRANSPLANTING ON RAISED BEDS OR PITS

Papaya plants are transplanted on 12–15 cm raised beds or raised pits to avoid direct contact of water with stem. It helps to prevent the spread of collar rot disease in papaya plants (Figure 4.6). It facilitates irrigation of orchards by flood system, easy fertiliser application, aeration of roots and weeding. The inter-row space of orchard can be easily filled by power tiller and this space may be utilised for suitable intercrops. The plants are earthed up easily and effectively before onset of the rainy season (Saran et al. 2014a).

5 Propagation and Layout

5.1 PROPAGATION

The papaya plants are propagated through asexual as well as sexual methods. In the case of asexual method, air-layering, cuttings and budding have been successfully practiced on a limited scale.

5.1.1 Air Layering

Air layering reproduces the characteristics of a preferred strain. All offshoots except the lowest one are girdled and layered after the parent plant has produced the first crop of fruits. Later, when the parent has grown too tall for convenient harvesting, the top is cut off and new buds in the crown are pricked off until offshoots from the trunk appear and develop over a period of 4–6 weeks. These are layered and removed and the trunk is cut off above the originally retained lowest sprout, which is then allowed to grow as the main stem. Thereafter, the layering of offshoots may be continued until the plant is exhausted. However, the aforementioned method is not feasible at a commercial scale due to the lower success rate.

5.1.2 Softwood Cuttings

Softwood cuttings made in midsummer (May and June) rooted quickly and fruited well. Terminal portions of the shoot should not be used unless the growth has matured and hardened. The cuttings can be cut into 10–15 cm (4–6″) length, and the leaves should be removed from the lower third of each cutting. Rooting hormones may be used to hasten rooting (Crocker 1994). Cuttings can be inserted into a rooting medium containing a 1:1 mixture of perlite and peat or other suitable materials such as vermiculite or sand. The medium should be well-drained and sterile. Cuttings taken in winters and spring are slow and deficient in root formation. Once rooted, the cuttings are planted in plastic bags and kept under mist for 10 days, and then put in a shade house for hardening before setting finally in the field. The longer (60–90 cm) cuttings are rooted more readily than smaller cuttings in the rainy season and they began fruiting in a few months very close to the ground. Cuttings are used for specific cultivars with parthenocarpic fruit (Bose and Mitra 1990). Rooting of cuttings is practiced in South Africa, especially to eliminate variability in certain clones so that their performance can be more accurately compared in evaluation studies.

5.1.3 Budding

Both forkert and chip methods of budding have proved satisfactory in Trinidad. However, it is reported that a vegetatively propagated selected strain deteriorates steadily and is worthless after three or four generations (Bose and Mitra 1990). 'Solo' grafted onto 'Dwarf Solo' was reduced in vigour and productivity, but 'Dwarf Solo' grafted onto 'Solo' showed improved performance. However, these techniques were not successful at the commercial level.

5.1.4 Tissue Culture

Papaya cultivation through seeds exhibits several problems due to inherent heterozygosity and dioecious (cross-pollination) nature of the crop (Bhattacharya and Khuspe 2001) and production of non-true-to-types (Panjaitan et al. 2007; Tsai et al. 2009). Seeds of open-pollinated flowers exhibit considerable variation in the shape, size and flavour (Drew and Smith 1986) and susceptibility to papaya ringspot virus (Clarindo et al. 2008; Tsai et al. 2009). Papaya multiplication through tissue culture has been started recently at the large scale with plants true to type and free from several diseases. The potential of rapid large-scale propagation of papaya selections by tissue culture is being explored and offers promises for the establishment of commercial plantations of superior strains. Efficient micro-propagation of papaya has become crucial for the multiplication of specific sex types and in the application of genetic transformation technologies (Lai et al. 2000). Significant progress has been achieved using organogenesis and somatic embryogenesis (Bhattacharya et al. 2003; Yu et al. 2003; Cabral et al. 2008). Moreover, in all these cases, shoot development was reported to have been accompanied by intervening callus phase. Callus formation is an undesirable feature during micro-propagation which leads to genetic variability of regenerated plants (D'Amato 1977). Using the epicotyl segments of a papaya cultivar, CO-7 as explants, an efficient multiple shoot induction was achieved via direct oraganogenesis from *in vitro* grown papaya seedling plants. This procedure generates only true to type plantlets, which could be extended for other papaya cultivars (Anandan et al. 2011).

Shoot bud induction occurred on culturing epicotyl segments in MS (Murashige and Skoog 1962) medium, followed by transfer onto shoot multiplication, elongation and rooting medium. Adventitious shoot induction in papaya was highly influenced by both of cytokinin and explant type (Liu et al. 2003; Ahmad and Anis 2007). TDZ (thiodiazuron) is extensively used for the induction of shoot regeneration in several plant species. The highest percentage of explants (80.50%) producing multiple shoots (6.3) was observed with 2.5 μM TDZ. A decrease in the number of shoots was noticed when the concentration of TDZ was increased from 2.5 to 10.0 μM. The lowest percentage of explants (34.75) producing shoots (1.4) was observed with highest concentration of TDZ (10.0 μM). Husain et al. (2007) have also supported the aforementioned findings that frequency of shoot regeneration declined markedly at higher concentrations of TDZ and was invariably associated with thick and stunted shoots. The frequency of shoot organogenesis, number of shoots or explants was found to vary significantly depending on the concentration of TDZ, as well as different time periods. However, the best concentration of TDZ (2.5 μM) produced highest (6.3) shoots after 6 week of culturing.

The explants with initiated shoot buds produced multiple shoots in MS medium supplemented with 5.0 μM 6-benzylaminopurine (BAP) and 0.05 μM naphthalenacetic acid (NAA) (Ahmad and Anis 2007; Anandan et al. 2011). Combination of BAP and NAA (5.0 μM each) was found best medium composition and gave optimum response for shoot induction along with the highest number of shoots or explants. Maximum shoot elongation was observed at 1.5 μM gibberlic acid (GA_3) concentration. Induction of rooting was found to be an extremely difficult process in papaya. Hence, different hormonal combinations were tried to find a suitable media for rooting. Half-strength MS medium without any growth regulator (control) failed to induce root formation even after 4 week. Elongated (60.25%) shoots were rooted on MS medium containing 2.5 μM indole-3-butyric acid (IBA) within 3 weeks after sub-culturing (Anandan et al. 2011). The *in vitro*-raised plantlets were successfully hardened in greenhouse with 72% survival rate (Figure 5.1). Under Indian conditions, this technique failed due to lack of properly standardised techniques, quality laboratory facilities and high cost of production. Due to which the farmers are not getting tissue-cultured plants of papaya today.

5.1.5 Conventional Propagation

Even though scion grafting and rooting of cuttings (Allan and MacMillan 1991) are possible, these methods are not routinely used for commercial papaya propagation. Propagation of papaya is mostly through seeds. Farmers, generally, collect fruits of good quality from their orchards and extract seeds for subsequent plantings. Numerous black seeds, enclosed in a gelatinous aril, are attached to the wall of the ovary in five rows.

The papaya seeds are very costly as they are produced under controlled conditions to maintain their genetic purity. Though, lots of problems are associated with seeds like, poor seed germination, poor seedling vigour, etc. under *ex-situ* condition, yet papaya seeds possess the highest viability and vigour at physiological maturity. Seed quality deterioration sets in early after extraction, resulting in low germination and seedling vigour. The subsequent reduction in seed germination after storage is also very fast due to accumulation of inhibitors in sarcotesta (Rayes et al. 1980). Higher the moisture content of the seeds along with high temperature of storage environment, the quicker is the loss of viability. Seed ageing causes gradual decline in all vital cellular components thereby, causing progressive loss of viability. The lipid auto-oxidation has also been suggested to be one of the causes of seed ageing. Seed deterioration leads to reduction in seed quality, performance and stand establishment. The priming with 1000 ppm GA_3 and 1 M KNO_3 in fresh and accelerated aged seeds showed significant improvement in seed quality parameters. Papaya seeds germinated poorly at 25°C in the presence of gibberellin (GA_{4+7}) or following matriconditioning at 25°C for 4 days. However, a combined treatment of air-conditioning and GA_{4+7} for 4 days synergistically promoted germination and seedling emergence. Drying the seeds after conditioning reduced the percentage of seedling emergence in the combined treatment involving 400 μM GA_{4+7} only. Combining matriconditioning with 100 or 200 μM GA_{4+7} could effectively reduce germination time, improve seedling emergence and is recommended as a standard procedure for testing papaya seed germination. The seedling vigour index was significantly higher in seeds treated with GA_3 at 100 ppm for 36 h whereas lower in

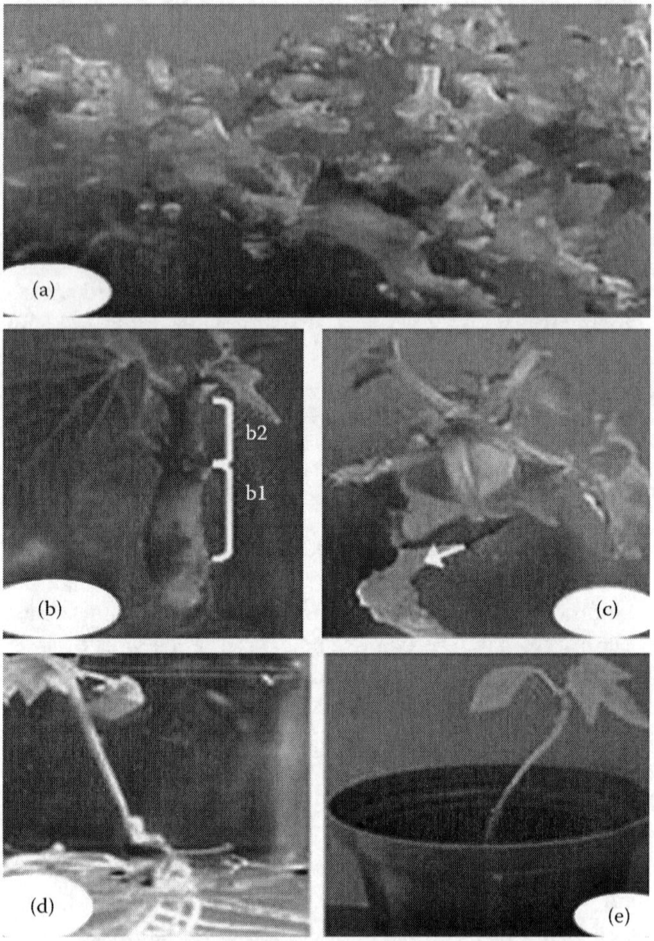

FIGURE 5.1 (**See colour insert.**) (a) Multiple shoot induction on MS medium with BAP (5 μM) + NAA (0.05 μM), (b) elongation on half MS with 1.5 μM of GA_3; (bl) stunted shoot before GA_3 treatment, (b2) elongated shoot after GA_3 treatment, (c) rooting of *in vitro* regenerated shoots in half MS with 2.5 μM of IBA, (d) plant with well-developed root system ready for hardening, (e) plant acclimatised to greenhouse conditions. (Adapted from Anandan, R, S. Thirugnanakumar, D. Sudhakar and P. Balasubramanian, 2011, *Journal of Agriculture Technology,* 7(5): 1339–48.)

untreated control with 48 h water soaking (Dhinesh Babu et al. 2010). Field germination of papaya seeds can be greatly improved by soaking them before planting in an aqueous solution of potassium nitrate (KNO_3). Germination percentage of both fresh and dried seeds can be increased with this method. The time to germination after the soaking treatment is reduced, and maximum germination is achieved sooner when untreated seeds are sown. Seeds soaked in KNO_3 solution produce seedlings that are initially more vigorous than seedlings from the untreated seeds (Yogananda et al. 2004).

5.2 RAISED BED SOWING

The farmers do not have infrastructure facilities like, poly-house, greenhouse, etc. for raising of seedlings under protected conditions. They grow the seedlings in field conditions. Seeds are usually treated with 0.1% thiram/cerasin/vitavax and planted in beds 15 cm above ground level that have been organically enriched and fumigated. The seedlings can be raised in nursery beds (300 cm × 100 cm × 10–15 cm) as well as in pots or polythene bags. The seed rate is 250–300 g/ha in gynodioecious cultivars and 300–500 g/ha in dioecious cultivars (Ram 1986b). The seeds are sown 1–2 cm deep in rows 10 cm apart and covered with fine compost or leaf mould or *khar* (dried leaves of *Saccharum*) or polythene sheets or dry paddy straw to protect and provide shade or maintain the moisture under Bihar conditions (Figure 5.2a). The shade must be removed soon after germination because papaya plants develop poorly if kept shaded. Light irrigation is provided during the morning hours. Germination may take 10–12 days during summer and rainy season, while during February to March, it takes 25–30 days due to low temperature under North Eastern Plains Zone of India (Figure 5.2b).

Authors observed the effect of temperature on per cent germination and days taken for germination was evaluated in *C. papaya* cvs. Pune Selection-3 and Pusa Dwarf under agro-climatic conditions of North Bihar (Figure 5.3). The highest seed germination percentage was observed in the month of July (average, 96.8% and 93.0%), followed by August (average, 94% and 88.5%), whereas the lowest germination percentage was observed in the month of January (44.2% and 36.8%) in Pune Selection-3 and Pusa Dwarf, respectively. Minimum number of days (average, 5.17 and 6.33 days) was taken for germination in the month of July and the maximum number of days (average, 29.5 and 31 days) was taken in the month of January in both the cultivars, PS-3 and Pusa Dwarf, respectively. The average maximum germination and seedling height was observed at temperature 29.8°C and 29.5°C during July and August, respectively, whereas the minimum was found at temperature 14°C and 15.9°C during January and December, respectively. The suitable time for nursery sowing was July to August for autumn season crop under subtropical conditions of India.

Papaya seedlings are transplanted 40–45 days after seed sowing when they are 15–20 cm in height. Transplanting is best done in the evening or on cloudy or

FIGURE 5.2 **(See colour insert.)** Nursery raising: seed sowing (a), seedlings on raised beds (b) and seedlings in poly tubes (c).

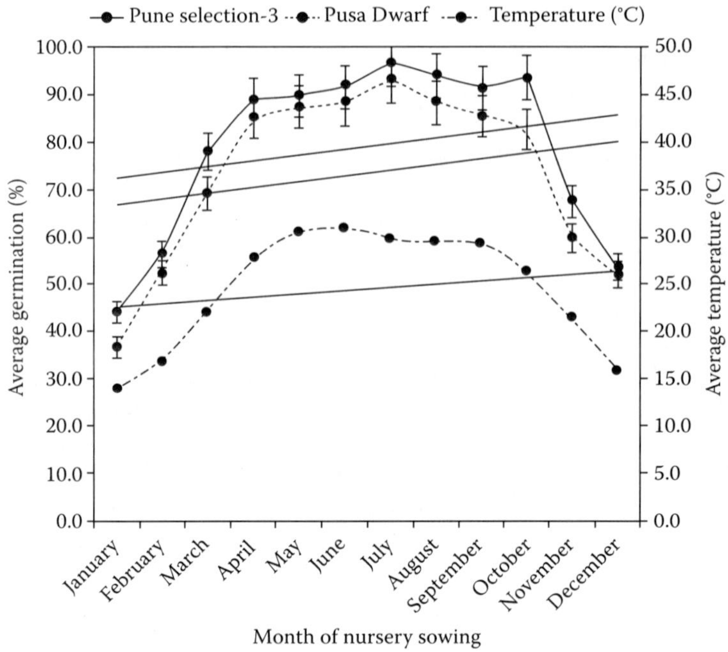

FIGURE 5.3 Relationship between average germination (%) and temperature (°C) in papaya cvs, Pusa Dwarf and Pune Selection-3.

damp days. On hot and dry days, each plant must be protected with a leafy branch or palm leaf stuck in the soil. Except hermaphrodite/gynodioecious varieties, the plants are set out in 10–15 cm apart in enriched pits. After flowering, one female or hermaphrodite plant is retained, the other two removed. At least one male is kept for every 10 females. In hermaphrodite and gynodioecious varieties such as Coorg Honey Dew, Surya, Solo, Red Lady, etc., one to two plants are required per pit because they are 100% fruiting plants. Watering is done every day until the plants are well-established, but over watering is detrimental to young plants.

5.3 PLASTIC BAG/DISPOSABLE GLASSES/ SOFT PLASTIC POT SOWING

Transplanting is more successful if polyethylene bags or disposable glasses filled with enriched soil are used instead of raised beds. Two to three seeds are planted in each bag but only the stronger seedling is maintained. Black coloured or transparent polythene bags of size (8–10 cm × 8 cm) are used with drainage hole for raising seedlings (Figure 5.2c).

5.3.1 Tray Sowing

The use of plastic tray for raising seedlings is a new way of seedling culture to obtain healthy seedlings and they are easy for transporting and transplanting. The tray size may be 74–82 holes at 4.5 cm each in diameter. Fill the well-prepared media in the holes, sow 1 to 2 seeds in each hole, and cover by a thin layer of the media. Seed would be sown directly in seedling trays or planting bags and thinned to the required number of plants per bag or it can be germinated in special containers and transplanted within 14 days after germination (two-leaf stage). The advantage of the latter is that the germination trays can be moved to a warm spot during the night. The number of seeds planted in a container will depend on the germination vigour and the cultivar. In general, three seeds are planted per container and plants are thinned as when required. For hermaphrodite cultivars, plant two seeds per container. Thinning is done in the field when the plants start flowering and sex may be identified. The extra plants should be transplanted into empty containers within 2 weeks after germination.

A good potting medium must drain well and be free of soil-borne diseases. Equal parts of a sandy soil and compost or equal parts of a loam soil, sand and compost can be used as a growth medium in the bags (Figure 5.4). If soil is used, sterilization is necessary. The growth medium should be analysed to correct deficiencies. The planting bags must be filled and irrigated well in advance of sowing. Any free water must drain away within 30 s after application. The medium usually contains adequate levels of phosphorus and potassium, but is low in available nitrogen. Thereafter, the seedlings are irrigated weekly using a balanced water-soluble fertiliser with high nitrogen content (DAIS 2009). It can also have a very low pH. Fertilisation and correcting of the pH may be done according to the results of the soil analysis. Fill the bags with the compost 2–4 weeks before sowing and irrigate regularly. Do not

FIGURE 5.4 Polytubes filled with potting medium for seed sowing.

sow the seeds in a dry medium because it will result in poor and uneven germination. For purposes of controlling aphid, viral infection, rain and wind protection and maintaining tolerable temperature during seedling stage, it is required to use screen house, greenhouse or tunnel covered with 0.07–0.10 mm plastic film. During the seedling stage, semi-humid environment is preferred. For better aeration, the film may be covered during the cool night or heavy rain period and opened during warm day time. The site of the bag, pot or tray should be changed if the roots of seedlings penetrate into the soil.

5.4 PLANTING TIME

In India, papaya is planted three times in a year, namely spring (February–March), monsoon (June–July) and autumn (October–November). In an experiment, it was observed that the highest yield and the lowest infection of virus were found when the seeds were sown in the nursery in 3rd week of August and planted in the field in the middle of October under subtropical conditions of North Bihar (Singh et al. 2010). These results are highly suitable for North Eastern regions of India including West Bengal, Eastern Uttar Pradesh, Bihar and Assam. The best time of papaya planting is during late summer (June–July) and autumn (September–October), except in areas where temperatures drop below 6°C in winter. In such areas, it should be planted from late spring to mid-summer (March–April) so that the plants are reasonably mature before the onset of winter.

The time of sowing depends upon the choice of fruiting season and danger of rain or frost. In the northern part of Taiwan, seeds are sown from March to May and transplanted from May to July. In central and south India, seeds are sown almost round the year, but optimum season is from February to March (Spring) or from September to November (Autumn).

Under agro-climatic conditions of South Africa, a late summer planting is ideal due to opportunity for plants to come to flowering phase during September to December. Plants transplanted in late spring to mid-summer will also begin to flower in September, but are much taller than those planted in late summer, leading to a shorter economic lifespan (DAIS 2009).

5.5 SITE SELECTION AND ORIENTATION

The three major environmental factors to be considered in selecting a site to grow papayas are temperature, moisture (rainfall and soil drainage) and wind. The hermaphrodite papaya plant preferred for commercial orchards is more sensitive to the growing environment than the female papaya plant, and therefore selection of a suitable site is critical. Another condition to consider is the amount of sunlight the site receives to support plant growth and fruit production. Insufficient sunlight results in low yields and fruits with inadequate sugar and encourages plant diseases affecting papaya production (Nishina et al. 2000). For optimal interception of sunlight, orchard rows should run in north or south direction.

5.6 LAND PREPARATION

First of all, the field should be levelled and then proper harrowing should be done with manuring and fertiliser incorporation. In a virgin field, the land is ploughed and harrowed twice, and then layout is done. An elevated plot along the row is made by ploughing on barring with two passes on each side. This will insure proper irrigation and drainage (Rex and Rivera 2005).

5.7 LAYOUT OF ORCHARDS

The producer often has to consider non-profitable male trees as part of the overall layout of an orchard. A growing demand for agricultural land, increasing land prices and production costs, all put pressure on the producer to increase the per hectare yield as well as the quality of the crop. There are various planting systems, namely square (Figure 5.5), rectangular, hexagonal, quincunx and paired row planting systems. Among them square and rectangular systems are very popular due to ease in layout with cultural practices. These can be adapted according to the spacing of the trees in the row. With these systems, adjacent trees within the row are not expected to touch one another even when they are fully grown and also allow for adequate space between rows for orchard implements. Pits of 50 cm × 50 cm × 50 cm size are dug about 2 month before planting. The first 15 cm of top soil should be kept aside and organic matter is added to improve the soil structure. The lower 15 cm of soil is discarded. Papaya plant needs heavy doses of manures and fertilisers. The pits are filled with top soil along with farmyard manure (20 kg), neem cake (1 kg), bone meal (1 kg) and carbofuran (50 g). Apart from the basal dose of manures applied in the pits, 200–250 g each of N, P_2O_5 and K_2O are recommended for getting high yield. Application of 200 g N is optimum for fruit yield, but papain yield increases

FIGURE 5.5 Layout and pit filling.

with increase in N up to 300 g. Limestone (about 450 g) is added for every unit rise in pH. Papaya requires a pH near neutral. The mixture is then added to the planting holes and built up into a mound; for best results this should be done about 6–8 weeks before planting. A 40–60 cm high bed is required if the soil is not well-drained.

5.8 PLANTING DISTANCE

Tall and vigorous varieties are planted at greater spacing, while medium and dwarf ones at closer spacing. Planting pit should be dug at a distance of 2.4 m × 2.4 m for tall, 1.8 m × 1.8 m for medium and 1.2 m × 1.2 m for dwarf type varieties. Normally, the distance between rows is about 2.0–2.5 and 2.0–3.0 m between plants. When plants are grown at sloppy land, the distance should be 3.0× 3.0 m; however, 2.7 × 2.7 m between rows for the mechanised farming practices is recommended (Table 5.1). The spacing of 1.4 m × 1.4 m or 1.4 m × 1.6 m is best suited for papaya cv., Pusa Delicious under subtropical condition of Bihar. A closer spacing of 1.2 m × 1.2 m for Pusa Nanha has been adopted for high-density planting accommodating 6400 plants yielding 103 tonnes of fruits per hectare (Singh et al. 2010a, b).

5.9 TRANSPLANTING

Make a slit in the plastic bag or disposable glass down one side to remove the bag and place the plant in the hole for proper transplanting of 45 days old seedlings. Do not break up and loosen the roots because this will cause the plants to lodge later. If there is more than one plant in a bag do not separate them for the same reason. Rake in the soil from around the hole to cover the roots and growth medium, but be careful not to pile too much soil around the stem of the plant. The surface of the growth

TABLE 5.1
Plants per Hectare in Different Papaya Planting Systems

Planting Distance (m^2)	No. of Plants/ha in Different Planting Systems		
	Square	Rectangular	Triangular
1.2 × 1.2	6400	–	–
1.8 × 1.8	3086	–	3571
1.8 × 2.1	–	2646	–
1.8 × 2.7	–	2058	–
2.1 × 2.1	2268	–	2625
2.1 × 2.4	–	1984	–
2.4 × 2.4	1736	–	2008
2.4 × 2.7	–	1543	–
2.7 × 2.7	1372	–	1585
2.7 × 3.0	–	1235	–
3.0 × 3.0	1111	–	1284

medium should be levelled with the soil surface. The young papaya plants should be planted in an upright position as far as possible. If not planted in this position, the tree will continue its slanting growth habit for the rest of its lifespan. The base of the stem of the adult tree will then lie on the ground and the upper two-thirds of the stem will form a flattened C shape, with the growing tip pointing straight upwards. As the tree ages, that is, grows older, its weight will cause larger part of the stem to lie on the ground, restricting movement in between rows. Rows need to be kept clear to facilitate weed control and harvesting. A young plant lying on its side is also more susceptible to fungal diseases, bird damage and sunburn.

5.10 AFTER CARE AND GAP FILLING

New transplanted seedlings should be taken care to protect against frost, water lodging, rainfall and pests. In subtropical India where winters are very cool, seedlings should be protected with polythene bags and small thatches. Some extra seedlings are kept in reserve for gap filling as per needs. The mechanical and chemical damage of young seedlings is frequently observed during weed management by cultural operations and spray of herbicides.

6 Nutrition

Papaya is a heavy feeder crop. Mineral nutrients are taken up by the plants grown under full sunlight as macronutrients (K > N > Ca > P > S > Mg) and micronutrients (Cl > Fe > Mn > Zn > B > Cu > Mo) for the proper growth and development of papaya plantation. Among them, nitrogen, phosphorus and potassium, which are very important in metabolism and frequently limiting in tropical soils, are extracted in high amount, a tonne of fresh-harvested fruits contain 1770, 200 and 2120 g of each of these nutrients, respectively, and among the micronutrients, namely, Fe, Mn, Zn, B, Cu and Mo are extracted 3364, 1847, 1385, 989, 300 and 8 mg of each nutrients, respectively. High-density plantings may extract 110, 10 and 103 kg of N, P and K per ha, respectively; however, this can be much higher, depending on the yield. The fruits signify 20%–30% of the nutrients removed.

If sufficient amount of balanced fertiliser is not applied to the plants, they will fail to produce fruit or give meagre fruiting. Adequate and efficient manuring of young and mature trees is essential to maintain the health of papaya and to obtain profitable yield. It starts from raising the seedlings in the nursery for the production of vigorous healthy plants. During preparation of nursery bed, 20.0 kg fine compost should be applied in each nursery bed measuring 3.0 × 1.0 m. After germination, where the seedlings are one month old, application of liquid solution of fertiliser ratio 10:10:10 produce healthy seedlings, especially in poor soils. Both organic and inorganic manures are beneficial to the papaya plants. Organic manures like, farmyard manure (FYM), sheep manure, neem cake, wood ash and bonemeal improve yield and quality of papaya fruits. Hence manuring in papaya should be half from the organic manures and half from the chemical fertilisers. Papaya responds well to the application of biofertilisers (*Azotobactor* and *Azospirillum*), and organic manures (vermicompost at 20 kg/plant) improve yield and quality of fruits.

An experiment in fruiting orchard, carried out all over the world, has indicated positive response to nitrogen, phosphorus, potash and micronutrients in papaya. Cunha and Haag (1980) estimated the removal of major and minor nutrients per tonne of papaya fruit. From the uptake studies, requirement of N, P and K was estimated to be 140:40:200 g/plant/year. However, from field experiments, 140–350 g N, 70–375 g P and 0–500 g K per plant per year have been found to optimum depending upon cultivars and agroclimatic condition. Application of 200 g N was also found optimum for fruit yield at Pusa but papain yield increased with dose up to 300 g at Coimbatore. Invariably 200–250 g each of NPK is recommended for high yield of papaya. The highest yield of Coorg Honey Dew papaya at NPK (250 g + 110 g + 415 g, respectively)/plant/year applied in six split applications. In another experiment, he reported a dose of NPK (125 g + 125 g + 500 g, respectively)/plant/year in six split applications to be the best. At Coimbatore, a dose of 200 g each of NPK/plant in four split doses during 1, 3, 5 and 7 months after planting resulted in higher yield of papaya variety, CO-1. For the papain production, a dose of 250 g

N/plant/year in six split doses at bimonthly intervals commencing from the second month after transplanting was found to be the best. Singh et al. (1998) found maximum fruit yield with 200 g N, 300 g P and 150 g K under Chhota Nagpur, Jharkhand (India) condition. A dose of 350 g N, 250 g P and 200 g K/plant/year applied in six split doses was the best for Solo variety spaced at 2 m × 2 m (2500 plants/ha) under Bangalore condition. A fertiliser dose of 200 g N, 300 g P and 600 g K/plant gave the highest fruit yield in Ranchi variety under West Bengal (India) condition. Optimum fertiliser dose (200 g N, 300 g P and 400 g K) for maximum papain yield (4.45 g/fruit) has been reported (Ram 2005). The following dose of manures and fertilisers/plant has been standardised after experimentation for several years at this station to obtain the maximum fruit yield under North Bihar (India). The maximum plant height (144.11 cm) was noted under the N_3 P_2 K_2 (300 g N and 250 g each of P and K/plant) which was followed by N_2 P_2 K_2 (250 g each of N, P and K). Flower initiation at maximum height (84.88 cm) was noted under the treatment of N_3 P_2 K_2 which was followed by N_3 P_1 K_2 (81.66 cm) but significantly at par with N_2 P_2 K_2. The maximum days (246.11) taken to flower initiation was N_2 P_2 K_2 treatment which was followed by N_3 P_2 K_2 and N_1 P_2 K_2 treatments which were statistically at par with each other (Rajbhar et al. 2010).

The doses listed in Table 6.1 makes a good balance between vegetative growth and fruiting. The above fertiliser dose should be applied in six split applications, once in two months commencing from the second month of planting. The fertilisers should be applied 20–30 cm away from the main stem and covered through earthing up or mixed in soil around the stem (Figure 6.1). The orchard should be irrigated lightly after the fertiliser application. The fertiliser may also be put in the irrigation ring and mixed thoroughly with the soil. Application of excess nitrogen, which is a common practice among the farmers, should be avoided as it increases vegetative growth and leads to poor fruiting.

Healthy well-fed plants that are adapted to the site, and part of a balanced plantation system that includes a wide variety of nutrients, will effectively preclude any economic losses of papaya production. Diagnosis will depend on good leaf-sampling technique with proper utilisation of soil and foliar nutrients, and good fertilisation practices. Before establishing a papaya plantation, a representative soil sample must be taken. The soil analysis (chemical composition) results will indicate the types and quantities of fertilisation needed before planting. An auger and two buckets can be used for the top and subsoil, respectively. A sample consists of a combination of

TABLE 6.1
Recommended Doses of Manures and Fertilisers for Papaya Cultivation

Organic Manures (Basal Dose)	Amount (kg)	Inorganic Fertilisers (Top Dressing)	Amount (g)
Compost	20.0	Nitrogen	200.0–250.0
Cake	1.0	Phosphorus	200.0–250.0
Bonemeal	1.0	Potassium	250.0–500.0

FIGURE 6.1 **(See colour insert.)** Fertiliser application in papaya.

at least 10 samples from 3 ha area. The NEPZ soils are deficient in boron, zinc and sulphur. The critical level of different nutrients in the soil of north Bihar is given in Table 6.2.

A sample provides information on the uptake of nutrients by the plant. Analysis of the petiole of the youngest fully expanded mature leaf beneath the most recently opened flower during November is carried out. Approximately 10–20 healthy plants are selected by walking diagonally through the plantation. The plants should be homogeneous in appearance and representative of the plantation. Samples should be collected in the morning after the dew has dried off. Samples should not be taken if plants are under stress as a result of drought, disease or high temperature. After heavy rainfall, wait at least for two weeks before taking samples. Normal nutrient levels in petioles are given in Table 6.3.

TABLE 6.2
Critical Nutrient Levels in Soils of North Bihar (India)

Sampling Year	Organic Carbon (%)	Available P_2O_5 (kg/ha)	Available K_2O (kg/ha)	Zn (ppm)	Cu (ppm)	Fe (ppm)	Mn (ppm)	B (ppm)	S (ppm)
2003	0.39	22.85	152	0.29	1.83	10.79	4.98	0.12	3.13
2004	0.59	27.42	169	0.38	1.98	11.54	5.19	0.24	1.56
2005	0.41	9.14	142	0.26	2.18	12.36	4.30	0.20	4.00

TABLE 6.3
Critical Nutrient Levels in Petiole of Papaya

Nutrient	Concentration (%)	Nutrient	Concentration (ppm)
Nitrogen	1.20–1.38	Iron	20–100
Phosphorus	0.17–0.21	Manganese	20–150
Potassium	2.70–3.40	Zinc	14–40
Calcium	1.00–3.00	Copper	4–10
Magnesium	0.40–1.20	Boron	20–50
Sulfur	0.30–0.80		

The imbalanced application of nitrogenous fertilisers may induce boron deficiency through crop growth dilution factors. Not many field trials have been conducted to determine the micronutrient requirement of papaya. The boron deficiency develops bunchy top and deformed fruit (Figure 6.2).

The ameliorative effect of boron application (as disodium octaborate tetrahydrate) through foliar sprays is more effective than soil application (1–3 kg/ha). However, it can start from 1 kg/ha for clayish soils, 1.5 kg/ha for rocky soils and 2–3 kg/ha for sandy soils. Foliar application of the micronutrients, namely, $ZnSO_4$ (0.5%) and H_2BO_3 (0.1%) is done in order to increase the growth and yield. It will also help in overcoming the deficiency. It is important to go for only one spray of boron in initial fruit development stage, if both are combined, it may prove toxic to the plant (Figure 6.3). Therefore, balanced application of macronutrients is imperative for quality production of papaya in boron deficient soils. The basal application of borax at 5 g/plant at the onset of reproductive stage is very effective to control the bumpy fruit disorder of papaya (Saran et al. 2014a, b, c, d).

FIGURE 6.2 (**See colour insert.**) Bumpy fruits of papaya.

FIGURE 6.3 (**See colour insert.**) Symptoms of boron toxicity on the leaves.

Ghanta et al. (1992) conducted an experiment to study the effect of foliar application of micronutrients, namely, B (0.1%), Mn (0.25%) and Cu (0.25%) applied singly and in combination at 2 and 3 months after transplanting on growth, flowering, yield and quality of papaya cv. Ranchi. All the micronutrients significantly increased the growth of plant, numbers of leaves produced per plant and length of petiole (5th leaf). Application of the micronutrients hastened flowering by 2–10 days. Combined application of Mn and Ca produced the highest yield, 97.44 t/ha compared with 67.01 t/ha in control. All the micronutrients increased individual fruit weight, fruit size and pulp thickness of fruit and reduced the seed content of fruit. Combined spraying of Mn and Cu showed best response in increasing the T.S.S. (9.67%) total sugar (6.84%), reducing sugar (6.29%), sugar/acid ratio (58:46), ascorbic acid (56.79 mg/100 g pulp) and total carotene (46.36 mg/100 g dry pulp) content of fruit.

The effects of mineral and elemental uptake of fertilisers by the plants are, however, significant and variable. High calcium uptake in fruit has been shown to reduce respiration rates and ethylene production, delay ripening, increase firmness and reduce the incidence of physiological disorders and decay, all of which result in increased shelf life.

7 Irrigation

Maintenance of optimum soil moisture is essential for plant growth, yield and quality of fruits. Once a plant is properly established, it grows continuously during winter months. It sets flower and fruit at every leaf axil that is produced. In general, 1.0 m^2 of leaves transpires 1.0 L H_2O daily, but can rise substantially with increasing evaporative demands. A papaya plant with 35 leaves, equivalent to approximately 3.5–4.0 m^2 leaf area, can fix ca. 70 g of CO_2 and transpire ca. 10 L of water daily (Ming and Moore 2014). For well-watered papaya plants, the crop irrigation coefficient (K_c) is close to unity but may reach values of 1.2, as a result of the strong dependence of canopy gas exchange, photosynthesis, water use and the solar radiation available (Campostrini and Glenn 2007).

Yield is obviously related to the number of leaves produced and the amount of growth added. Hence, water is required during the growing period for potential yield. Under low moisture conditions, floral sex shifts towards female sterility resulting in low productivity (Ram et al. 1985a, b). Responses to water stress include dehydration postponement through strict stomatal regulation, cavitation repair and intense osmotic adjustment (Mahouachi et al. 2006). At the same time, over-irrigation may cause root rot diseases. Thus, efficient water management is required. Number of irrigations depends upon soil type and weather conditions of the region. Protective irrigation is required in the first year of planting (Figure 7.1a). In general, irrigation to grown-up plants is given once in 7–10 days in winter and 4–5 days in summer. In the second year, when the plants are ladden with fruits, irrigation at fortnightly interval in winter and at 10 days interval in summer is needed from October to May. Excessive moisture is more detrimental to plant than moisture stress necessitating the effective drainage system, especially in heavy soils under high rainfall conditions, to avoid plant mortality. Studies conducted at Coimbatore have also indicated that papaya has high tolerance for heat and soil moisture for higher productivity; however, moisture stress at fruit development stage should be avoided. Total water requirement of CO-2 papaya under tropics is estimated to be 1800–1900 mm and excessive depletion of moisture causes reduced growth and yield. However, intermittent moisture stress induces deep root penetration. Depletion of soil moisture also causes increase in N and Mg content, while Ca content is reduced. Irrigation at 60% available soil moisture depletion is found to be the optimum for papaya. Regular irrigation is an important aspect in papaya cultivation, which helps in the growth, fruit development and high yield. Moisture stress inhibits the growth and also promotes male floral characters. The ring system of irrigation has been found very effective among different methods. This system helps in preventing the collar rot (Figure 7.1b) as there is no direct contact between the water and the stem portion. In low rainfall areas, where water is scarce, sprinkler or drip systems of irrigation can be adopted for higher production, saving water from 50 to 60%.

FIGURE 7.1 Protection of seedlings from frost under ridge system (a), ridge and furrow system (b), drip irrigation (from anubhavagro.tradeindia.com) (c), and conventional method in papaya (d).

Micro irrigation is one of the latest innovations for applying water and it represents a definite advancement in irrigation technology. It can be defined as a frequent application of small quantity of water directly, above or below the soil surface; usually as discrete drops, continuous drops, tiny streams or as miniature sprays through mechanical devices called emitters or applicators, located at selected points along the water delivery line. Types of micro irrigation systems include surface drip, subsurface drip, spray irrigation and bubbler irrigation. The surface drip includes both online drip system and integral drip line system. The irrigation schedule is fixed on the basis of soil type and weather conditions of the region (Saran et al. 2013a). In papaya, the increase in yield with the adoption of drip varied from 30 to 75% over surface methods of irrigation. On the basis of three years of trials, it is suggested to adopt drip irrigation (Figure 7.1c) with mulching in papaya (Pusa Dwarf) with 54.24% increase in the yield in comparison to the traditional system of irrigation without mulching (Saran et al. 2013a). Drip irrigation with 8 L water/plant/day coupled with plastic mulching led to increased fruit yield (83.3 fruits/tree compared to 74.6 fruits/tree in control) and size (1.8–3.2 kg compared to 1.3–2.4 kg in control). The study revealed that drip irrigation at 60% volume of available soil moisture (ASM) or water content at field capacity with black plastic mulch of 25 μm thickness resulted in higher yield (1264 q/ha) and yield attributing characters as compared to other treatments. The net seasonal income obtained was Rs. 251,962/–. The B/C ratio was found to be the highest (1:9.7) under drip irrigation with 60% volume of water and plastic mulch (Chaterlan et al. 2012). Application of irrigation water through drip system in papaya (cv. Red Lady) is economical and cost-effective over the conventional (basin) irrigation method (Figure 7.1d). Earthen up the papaya plants which helps in preventing the plants from lodging against windy strokes and also irrigate through bed system of flooding. Water use efficiency is increased by adopting drip irrigation. Through the drip system, 6–8 L water/plant/day is required for better yields. For environmental and irrigation management purposes, it is important to accurately estimate its crop evapotranspiration and irrigation requirements.

7.1 DRAINAGE

Papaya needs frequent but light rainfall or irrigation, and it must have good drainage. It is recommended that the papaya plants are very susceptible to water logging. Flooding for 48 h is fatal and it may kill the well-established orchard (Ram 2005). The succulent roots do not tolerate excess water, and two days with hypoxia causes chlorosis, leaf shedding and even death after 3–4 days with oxygen deprivation (Campostrini and Glenn 2007). It is, therefore, most important to select upland for papaya plantation. It may further be shaped sloppy in heavy rainfall areas. It is essential to make few furrows or trenches for quick and complete drainage of water during rainy season. Earthing up around the plants may be repeated more than four times during fertiliser application before the onset of monsoon. Under heavy irrigation, papayas grow optimally in the soils with an unimpeded depth of more than 1.0 m. However, if irrigation is well-planned and managed, there should be no problem of soil with an unimpeded depth of 75 cm, provided that no drainage problems occur at that depth (Saran et al. 2013a).

8 Seed Production

8.1 PLANT TYPE

The papaya seed production requires technical knowledge due to its complex genetic make-up and cross-pollinated nature. The monoecious papaya needs cross-fertilisation, and therefore, segregates very rapidly when male plants of different varieties are found at one site. In order to harvest genetically pure seeds, it is necessary to ensure varietal purity. This is achieved by removing any unidentified male plant in the proximity of the plantation. When seed results from open pollination, it becomes impossible, in most cases, to obtain selections which are reasonably uniform in flower type and fruit characteristics. Depending upon the sex type and parental combination, papaya varieties fall under two broad groups as mentioned below.

8.1.1 Dioecious Varieties

Varieties which produce male and female plants separately are called dioecious. The proportion of male and female is generally 1:1 with male population slightly on higher side which is a genetical character. The prevalent varieties which fall under this group are Pusa Dwarf, Pusa Nanha, CO-1, CO-2, Washington, etc.

8.1.2 Gynodioecious Varieties

The gynodioecious varieties comprise female and hermaphrodite plants and both are productive, thus, these varieties possess only productive plants and are preferred because of the absence of unproductive male plants. The varieties coming under this group are Pusa Delicious, Pusa Majesty, Coorg Honeydew, Solo (Hawaii), etc.

8.2 SEED PRODUCTION METHOD

The seeds in dioecious varieties are produced by pollinating the female flowers with male flowers of the same variety, known as sib-mating. The seed in gynodioecious varieties are produced by crossing the female flower with the hermaphrodite. Since variation in sex in hermaphrodite plants occurs under different environmental conditions, selection of regular and prolific bearing hermaphrodite plant(s) for seed production in each generation is essential. In gynodioecious varieties, the proportion of female and hermaphrodite plants will depend upon the technique used. If the female plant is crossed with the pollen of hermaphrodite plant, the population of females and hermaphrodites will be in equal proportion. If the hermaphrodite plant is selfed or intercrossed, the proportion of female and hermaphrodite in the plant population will be 1:2. The female plant is preferred because of the higher yield. Hence, the seed is produced by crossing the female flower with the pollen of a hermaphrodite plant in gynodioecious varieties. The yield variation in hermaphrodite plants is very high and, as a result, this plant is not preferred (Ram 1995).

The seed production in papaya with 100% genetic purity is difficult because of the dioecious nature of the plant. Therefore, the seed should be produced either strictly under controlled conditions or in an isolated area. The breeder seed is, generally, produced under controlled conditions. Under this method, the stigma of female flowers is brushed with the pollen of male flower of the same variety in dioecious varieties. While crossing female flowers, care should be taken that stigma is not damaged. Only those flowers which are ready to open should be selected in both the parents (Saran et al. 2014c). This avoids advance covering of flowers against contamination with foreign pollen. Similarly, for producing seed in gynodioecious varieties, the ready to open female flower should be crossed with the pollen of hermaphrodite flowers/plant of the specified varieties in the same way. Seed production by selfing hermaphrodite is much easier because the selected flowers are only covered with glassine wax/butter paper. After crossing and selfing, it should be properly tagged for identification. Before starting seed production of gynodioecious variety, knowledge of different types of flowers produced by a hermaphrodite plant is essential. Generally, a hermaphrodite plant produces five different types of flowers simultaneously; these are reduced elongate, elongate, carpelloid elongate, carpelloid pentandria and pentandria. Except reduced elongate, all the four types of flower are bisexual. The reduced elongate is unisexual and produces only pollen. The fruit shape of hermaphrodite plant depends upon the type of flower produced. The elongate flower produces a cylindrical- or cucumber-shaped fruits while carpelloid elongate produces deshaped or cat-faced fruit. The seed contents in these fruits are comparatively low. The pentandria and carpelloid pentandria produce round to round oval-shaped fruits which resemble the fruits of female plant.

Directives:

- Select the true to type trees (colour, yield, fruit shape, dioecious and hermaphrodite) in the plantation of particular variety.
- Cover the flowers with a bag once they form to ensure self-pollination in hermaphrodite flowers.
- Cover the flower buds with a bag once they start changing colour from light green to yellowish on tip and just before opening or slitting in dioecious.
- Similarly, the male flowers should also be covered before these open.
- Next day, carefully brushes the pollen collected from the covered male bud or mature bud (just before opening) on female flower.
- Mark these pollinated flowers.
- Fruits are collected at maturity or colour break stage.
- In breeding or seed production programmes of dioecious varieties, the male plants are been tagged on the basis of ideotype/desired characters, namely, flower colour, size, morphology, plant height, peduncle/petiole length, etc.
- If desired male plant has been tagged for different traits, it should be maintained for more than one year.

8.2.1 Seed Production in Isolation

Under controlled conditions, it is possible to produce seed with 100% genetic purity but it is a cumbersome job and may not be feasible for the majority of the growers. Moreover, it requires manual labour with some technical knowledge and experience. Therefore, seed can also be produced in isolation which does not require any elaborate process in crossing and selfing. The foundation and certified seeds are generally produced following this method. Isolation distance is an important factor for pure seed production of different cultivars (Saran et al. 2014a). The isolation distance may be kept at 1000–1500 m (Ram 1995). This distance depends upon the locality and activity of vectors. For ensuring a definite demarcation line of isolation in a locality, the simple method is transplanting of Washington or Homestead variety as a gene marker. The purple colour of stem and petiole is a dominating character in this variety. If the progeny of open pollinated plants do not produce any purple-coloured plant, it means the isolation distance is correct. This distance may be further increased if there are some purple-coloured plants found in the progeny. If a suitable isolation is not available, seed can be produced in the middle of an orchard of any crop like mango, litchi, sapota, ber or chestnut (Ram 1986a). These fruit trees act as a physical barrier against cross-pollination with other varieties of papaya in a locality. The seed in isolation generally produces 95%–99% pure seed depending upon the local condition. The entire fruit of the orchard can be utilised for seed extraction. This can be done where the market is available for cut fruits/pulp or a factory for the processing purpose. The planting season may start from March to October in different climatic conditions.

Sometimes, collar rot is noticed. The plants planted on raised beds or ridges help to avoid the disease. It may be controlled with repeated sprays of Bordeaux mixture (5:5:26). In areas susceptible to root rot, application of 1.0 kg lime and 100 g copper sulphate in the pits is an effective preventive measure against this disease.

8.2.2 Seed Extraction and Storage

The fruits should be harvested when they start ripening or reach colour turning stage (2%–4% ripening). Cut the fruits longitudinally and collect the seed carefully in a container with the help of spoon or knife (Figure 8.1). Fresh seeds put overnight for fermentation followed by sarcotesta should be removed by rubbing against gunny bags and finally washed in running tap water. After this, the seeds are dried in shade till the moisture comes down to 8%–9%. After drying, the seeds should be cleaned properly by removing the immature seed and other foreign materials. The standard method of storing papaya seed is to keep it in a refrigerator at 5°C. However, papaya seeds are sensitive to low temperature storage. Lipid crystallisation occurs at temperatures below –13°C to –15°C and seeds suffer damage if subsequently imbibed without briefly raising seeds to around 45°C to allow lipids to safely melt. The species-specific critical temperature seems to be related to triacyglycerol composition. This facility may not be available everywhere. Therefore, the seed should be kept in airtight bottles or packed in polythene bags and dry place under ordinary room temperature (Ram 1995). A standard

FIGURE 8.1 (**See colour insert.**) Traditional method of seed extraction.

has been fixed for the germination of papaya seed. The freshly extracted seeds give 90%–100% germination. But as the time passes, the germination percentage decreases. The stored seed supplied to growers should not have less than 70% germination in gynodioecious varieties and 60% germination in dioecious varieties (Ram 1986a). Seed treatments to promote germination and to reduce germination time have been widely investigated by Salomao and Mundim (2000). Satisfactory germination percentages were obtained by removing the sarcotesta, exposing dry seeds at 10°C prior to sowing, soaking seeds in distilled water, potassium nitrate, thiourea, sodium thiosulfate, tannic acid/ferulic acid and GA_3, and after that sown at 20–30°C temperatures.

8.3 SEED YIELD

The yield of papaya seed depends upon the cultural practices, sex ratio and pollination in tropical areas, while seed yield also depends upon the severity and duration of winter period in subtropical areas. Under normal conditions, the seed yields are 60–75 kg/ha in dioecious varieties. The seed yield is much lower in gynodioecious varieties because of the poor pollen production in hermaphrodite plants and, therefore, the cost of seed production is very high in these varieties. Despite high cost, farmers prefer the seed of gynodioecious varieties because of saving in raising crops or roguing of male plants (Ram 1986a).

8.4 CONSTRAINTS IN SEED PRODUCTION

Papaya is a remunerative but problematic crop and is very much sensitive to agro-ecological and other factors. Several factors responsible for losses to the farmers and seed producer in the country are given below.

8.4.1 Seed Viability

Papaya seed survives for a short period of time under ambient conditions. The seeds show an orthodox storage behaviour (Ellis 1984) in which they withstand desiccation and extend longevity at lower temperatures. Papaya seeds are susceptible to chilling temperatures and are killed when stored at zero or subzero temperatures. Therefore, the seeds are considered intermediate between recalcitrant and orthodox attributes. Seed moisture content is very crucial in papaya seed storage. Higher seed moisture content injures the seed life and rapidly reduces seed viability during storage. Harrington (1972) found that papaya seed longevity decreases by one-half for every one percent rise in seed moisture content, and that it well maintains at the range of 5%–14% moisture content. Seed moisture of 5%–7% is ideal for safe storage of papaya seeds. In general, the moisture content of seed increases with a rise in relative humidity level, and absorption depends on the surrounding temperatures. Papaya seeds deteriorate rapidly at higher storage temperatures and relative humidity. Fresh seeds give higher germination rate and seedling vigour that will decline with increase in the storage time. Zulhisyam et al. (2013) reported that the seeds with 6%, 8% and 10% moisture contents using silica gel are stored at 0°C, 4°C and 28°C, respectively, for three months. Seeds containing 6% moisture content and stored at 0°C gave higher percentage of germination, lower dormancy and lower seed death compared to the seed of the other storage conditions. The result suggested that such a condition was the best for papaya seed storage. Seeds containing 10% moisture content and stored at 28°C is not recommended for papaya seed storage because seed deterioration rate under such a condition was higher.

The effect of seed moisture, packaging materials and storage temperature on seed viability during storage was investigated in three varieties of papaya. The seeds were dried to various moisture levels, packed in polythene cover (700 gauge), aluminium poly pouches and butter paper bags and stored under controlled (15°C and RH 30%) and ambient conditions. The observations on seed germination and seedling vigour were recorded at 3-month intervals for 24 months. The results revealed that the seeds could be stored without any decline in viability and vigour for 24 months at 15°C irrespective of variety, packaging materials and seed moisture levels. However, under ambient conditions, the seeds of Surya could maintain high viability upto 15 months of storage, whereas Co-2 and Co-7 could maintain viability (>75%) upto 24 months with little decline in seed vigour. Among the packaging materials, poly-lined aluminium pouches were found to be better and the seed moisture (5%–10%) had no marked effect on viability (Yogeesha et al. 2008).

The seed is enclosed in a gelatinous sarcotesta (aril or outer seed coat), which is formed from the outer integument. Papaya seed germination is slow, erratic and incomplete. While the sarcotesta can delay germination, dormancy is also observed in seeds from which the aril has been removed. Several attempts have been made to overcome dormancy and improve papaya seed germination. Treatments such as removing sarcotesta, presoaking in water or water leaching promote *Carica* spp. germination. Seeds germinated poorly at 25°C in the presence of gibberellin (GA_{4+7}) or following matriconditioning at 25°C for four days. However, a combined treatment of matriconditioning and GA_{4+7} for four days synergistically promoted germination and seedling emergence. Drying the seeds after conditioning reduced the percentage of seedling emergence in the combined treatment involving 400 μM GA_{4+7} only. Combining matriconditioning with 100 or 200 μM GA_{4+7} could effectively reduce germination time and improve seedling emergence, and is recommended as a standard procedure for testing papaya seed germination (Claudinei and Khan 1993).

8.4.2 Temperature

Papaya thrives best in areas where the maximum temperature does not exceed 40°C and the minimum does not drop below 10°C. The plants are very susceptible to cold and frost right from germination till maturity as these not only affect the plant but also disfigure the fruits due to oozing of latex. Temperature fluctuation or severe winters from December to February affect the seed production (Figure 8.2). On ripening, the frosted fruit do not have a normal taste and flavour and, thus, remain unmarketable. In the plains of North India, papaya crop is adversely affected during summers by the hot winds, thus, frequent irrigation is necessary during this period. Prolonged droughts associated with high temperatures adversely affect fruit production by inducing abortion of floral and fruit structure leading to sterile phases or skipping of fruiting along the stem. Due to high temperature during summer, vivipary also affects the seed production. Fruits grown in different states of India, namely, M.P., Rajasthan, Western U.P., Punjab, Haryana, Delhi, H.P. and Jammu are affected by either low or high temperature or both. In tropical region round the year, this crop may be taken economically as compared to subtropical areas due to severity and duration of winter period.

8.4.3 Sex

The sex forms of papaya present a special problem both to the breeder and to the grower. The gynodioecious varieties claimed to possess cent percent productive plants, a wide variation in hermaphrodite population is observed giving from a few fruits to a number of unmarketable fruits. Under such circumstances, farmers have to shift to dioecious varieties giving 50% productive plants. By keeping a population of 5%–10% male plants in a dioecious population, a grower should aim at securing

FIGURE 8.2 **(See colour insert.)** Deformed and discoloured seeds of papaya.

a maximum number of female plants (Saran et al. 2014c). This can be accomplished in two ways. Either he should be able to identify their sex in the nursery stage or he should plant at least three seedlings at each place during transplanting of papaya in the field. The latter technique involves thinning out the male plants at the first expressing of sex, a rather cumbersome method.

8.4.4 Non-Availability of Quality Seed

Seed is the most potential factor determining successful cultivation of papaya. Mostly the seeds are procured by majority of the farmers from nurseries of local market which have the mixture of different local types. The mixed seeds are being supplied to growers under different names. Under the prevailing conditions, the purity of seed is generally not reliable. Ironically, there is no organisation like, State Seed Corporation or National Seed Agency in the country which produces and distributes genuine papaya seed to growers. Only a few agricultural universities and research centres in the country are dealing with papaya.

8.5 SEED CERTIFICATION

The purpose of seed certification is to maintain and make availability to the public, through certification, high-quality seeds and propagating materials of notified kind and varieties so grown and distributed as to ensure genetic identity and genetic purity. Seed certification is also designed to achieve prescribed standards. Certification is conducted by the Certification Agency notified under Section 8.8 of the Seeds Act, 1966. Certified seed producer means a person/organisation that grows or distributes certified seed in accordance with the procedures and standards of the certification. Seed of only those varieties which are notified under Section 8.5 of the Seeds Act, 1966 shall be eligible for certification (Trivedi and Gunasekaran 2013).

8.5.1 Clone Certification Standards

The general seed certification standards are basic and together with the following specific standards constitute the standards for certification of papaya seed. Land to be used for clonal propagation of papaya shall be free from volunteer plants. A minimum of three inspections should be made, the first before flowering, the second during flowering and fruiting stage and the third at mature fruit stage for the harvesting of fruits for seed extraction (Trivedi and Gunasekaran 2013).

8.5.2 Field Standards

The clone propagation plots of papaya, isolated from the contaminants, are 1500 m and 1000 m under foundation and certified classes, respectively. A maximum limit of off type plants, 0.1% in foundation and 0.2% in certified classes at and after flowering, has been permitted. Other seed standards are given in Table 8.1 for both the classes.

TABLE 8.1
Seed Standards for Foundation and Certified Seed Classes

Seed Standards	Foundation	Certified
Pure seed (minimum)	98.0	98.0
Inert matter (maximum)	2.0	2.0
Other crop seeds (maximum)	None	None
Germination (minimum)	60.0%	60.0%
Moisture (maximum)	7.0%	7.0%
For vapour-proof containers (maximum)	6.0%	6.0%

Source: Adapted from Trivedi, R. K. and M. Gunasekaran, 2013, Indian minimum seed certification standards, The Central Seed Certification Board, Department of Agriculture & Co-operation Ministry of Agriculture, New Delhi: Government of India, p. 605.

TAG No.............. KIND.............. Variety.................... Lot No......................

Certification Agency's EMBLEM

Name & Address of Certification Agency

Certified Seed Class of seed..................... Certificate No.................. Date of issue of

Certificate....................... Date of test......................

"Use of the seed after expiry of the validity period by any person is entirely at his risk and the holder of the certificate shall not be responsible for any damage to the buyer of seed. No one should purchase the seed if seal or the certification tag has been tempered with"

Certificate valid up to.......................... (Provided seed is stored under cool and dry environment)

Validity of certificate further extended upto

Name and Full Address of the Certified Seed Producer..

N.B.: If tag is to be affixed on a smaller container then the size of the tag may be reduced proportionately. However, length and breadth ratio and contents would remain the same.

FIGURE 8.3 Contents and layout of different seed certification tags.

8.5.3 Specification for Certification Tag

Seed certification tag should be 15 cm in length and 7.5 cm in breadth, and is made of durable material such as thick paper, paper with cloth lining, wax coated paper, plastic coated paper, etc. Both sides of tag paper shall be white for foundation class and blue (ISI No.—Azure blue) for certified class. The detailed contents and layout of different seed certification tags are shown in Figure 8.3.

9 Pests and Birds

9.1 MEALY BUG

The papaya mealy bug, *Paracoccus marginatus* Williams and Granara de Willink (Hemiptera: Seudococcidae) is a small polyphagous sucking insect with pest status that attacks several genera of host plants, including economically important tropical fruits, vegetables and ornamentals.

9.1.1 Symptoms

Infestation of the mealy bug appears as clusters of cotton-like masses on the above-ground portion of plants with long waxy filaments (Figure 9.1). Immature and adult stages of *P. marginatus* suck the sap of the plant and weaken it. Curling, crinkling, twisting, distorted, size reduction and yellowishness began to occur in the leaves which ultimately wither. The honeydew excreted by the bug results in the formation of black sooty mould which interferes with photosynthesis and causes further damage to the crops. The insect sucks the sap by inserting its stylets into the epidermis of the leaf, fruit and stem. While feeding, it injects a toxic substance into the leaves, resulting in chlorosis, plant stunting, leaf deformation or crinkling, early leaf and fruit drop, and death of plants (Tanwar et al. 2010). Heavy infestations are capable of rendering fruit inedible due to the buildup of thick white waxy coating. Papaya mealy bugs are most active in warm and dry weather.

9.1.2 Management

Mealy bug control often involves the control of attendant ants that are important for the proper development of mealy bugs. Without the ants, mealy bug populations are small and slow to invade new areas and the field would be free from serious mealy bug infestation. Therefore, management of mealy bugs often includes the control of ant species (Miller and Miller 2002).

9.1.2.1 Cultural and Mechanical

Pruning, removal, burning of infested crop residues, removal of weeds/alternate host plants like avocado, cherry, pigeon pea, guava, acalypha, hot pepper, tomato, wax apple, eggplant, sugar apple, glyricidia, sweet potato, hibiscus, mango, madeira fig, acacia, plum, ginger lily, parthenium etc., in and around nearby crop (s) will be very effective. Avoiding the movement of planting material from infested areas to other areas and also avoiding flood irrigation restrict the spread of infestation. Prevention of the movement of ants, destruction of already existing ant colonies and sanitisation of farm equipments before moving it to the uninfested crop would be beneficial

FIGURE 9.1 (**See colour insert.**) Mealy bug infestation on papaya plants.

practices. Use of sticky bands or alkathene sheet or a band of insecticide paste on branches or on main stem to prevent movement of crawlers from ground to trees is an effective practice.

9.1.2.2 Biological

Natural enemies of the papaya mealy bug include the commercially available mealy bug destroyer *Cryptolaemus montrouzieri*, ladybird beetles, lacewings, *Scymnus* sp. and certain hymenopteran and dipteran parasitoids. Conservation of these natural enemies in nature plays important role in reducing the mealy bug population. In the nature, lepidopteran predator, *Spalgiepius* sp. (Lycaenidae) is a well-known representative of carnivorous butterfly feeding on ovisacs, nymphs and adult stages (Muniappan et al. 2006).

9.1.2.3 Chemical

Locate ant colonies and destroy them by drenching with chlorpyriphos 20 EC at 2.0 mL/L of water. Apply chemical insecticides as the last resort such as dimethoate 30 EC (2.0 mL/L), or imidacloprid 17.8 SL (0.6 mL/L) two times at 15 days interval. Spot application of insecticide immediately after noticing mealy bug is again very effective. If the activities of natural enemies are not observed, use of botanical insecticides such as neem oil (1.0%–2.0%), NSKE (5.0%) or fish oil *Rosin* soap (25.0 g/L of water) should be the first choice. Drenching soil with chlorpyriphos around the collar region of the plant to prevent movement of crawlers of mealy bug and the ant activity is useful (Tanwar et al. 2010). Use of chlorpyriphos should be done with caution as foliar spray on leaves causes leaf burning in papaya as observed by the author.

9.2 FRUIT FLY

Papaya fruit fly (*Toxotrypana curvicauda*) is the principal insect pest of *Caricaceae papaya* throughout the tropical and subtropical areas. The insect deposits its eggs in the papaya fruit. After about 12 days, the larvae emerge and feed on the developing seeds and internal portions of the fruit. Infested fruits subsequently turn yellow and eventually fall from trees prematurely (Mossler and Nesheim 2002). However, damage to the fruit is not the major problem affecting production, rather the fruits from regions with fruit flies are restricted for export to regions that do not have these pests, unless given a postharvest hot-water treatment (Reiger 2006), is considered as a major hurdle.

Fruit fly is a major pest of papaya as concerned with fruit export. The female deposits eggs in the mature fruit which will later be found infested with the larvae (Jang and Light 1991). Only thick-fleshed fruits are safe from this pest. These are sometimes called wasps, because of the long ovipositor of the female fly as well as similarities in size and colour (Figure 9.2a). This long egg-laying organ is as long as the body penetrates flesh of the fruit and finally enters in seed cavity. Eggs are usually laid in small fruits, about 2–3 in in diameter, but they may be deposited in smaller and larger fruit, especially during high populations of the fly. The larvae, which are small legless maggots, feed on the seed and interior parts of the fruit (Figure 9.2b). When the larvae become mature, they emerge from the fruit, drop to the ground beneath the plant and pupate just below the soil surface. After about two to four weeks, the flies emerge to mate and seek fruit to lay eggs (Pena and Johnson 2006).

9.2.1 MANAGEMENT

It is too late to attempt control measures after the female fruit fly has deposited eggs in the fruit. Consequently, control procedures should be directed for preventing egg-laying either by mechanical means or by applying insecticides to kill the adult

FIGURE 9.2 **(See colour insert.)** (a) Female fruit fly and (b) damaged fruit. (Adapted from Pena, J.E. and F.A. Johnson, 2006, *Insect Management in Papaya*, The Institute of Food and Agricultural Sciences (IFAS), University of Florida, ENY-414:1–4.)

female before she deposits her eggs. Imidacloprid (1.0 mL/3.0 L of water) is recommended. It may be applied 7–10 days before the harvest. Insecticides should not be applied more than 6 times per season (Pena and Johnson 2006). The population of fruit fly reaching 7 individuals/trap/week during monitoring by applying 1 trap/ha is considered to be alarming situation.

India has no fruit fly with ovipositor long enough to lay eggs inside papayas. Home gardeners often protect the fruit from attack by covering with paper bags, but this must be done early, soon after the flower parts have fallen, and the bags must be replaced every 10 days or 2 weeks as the fruits develop. Fruits are harvested at the mature green stage, and over-ripe and infested fruits should be disposed properly to avoid further spread of fruit fly.

Control of the fly may be achieved by mechanical protection such as the use of paper bags but it is very difficult to practice on a commercial scale. Each fruit may be enclosed in a 1.5–2.0 kg size bag tied around the fruit stem to hold the bag. Newspaper, one-half sheet (about 12–15 in in size), may be rolled to enclose the fruit, then tied around the fruit stem, and also the free end. Bagging should begin when the fruit is small, shortly after the flower parts have fallen. This method of control is more adapted to small (1–25 plants) area than to a large (one-fourth acre or more) area. Although bagging the fruit is the most certain method of control, it is a labourious process that requires attention at regular intervals (10–14 days) to keep the young fruit covered. Also, this procedure will injure some of the fruit unless handled carefully. Sanitation is important in the control of the papaya fruit fly. It consists of destroying all dropped and prematurely ripe fruits, as well as small fruits suspected of being infested to prevent the larvae from developing into adult fruit flies (Pena and Johnson 2006). However, this is not a feasible practice on large commercial orchards since it is a labourious procedure, requiring regular monitoring and fruits can easily be damaged, unless handled carefully. Feasibility of using parasitic wasps as biocontrol agents is under test (Nishina et al. 2000).

Papayas must be treated before export to avoid introduction of fruit flies in new areas. Fruits picked 1/4 ripe are prewarmed in water at 43°C for about 40 min then quickly immersed for 20 min at 48°C. This double-dipping may be replaced by irradiation. Fruits that had hot water treatment followed by irradiation at 75–100 krad help to overcome the fruit fly. Use of different traps available in the market based on sex pheromone (2-methyl-6-vinylpyrazine) produced by male to avoid use of hazardous chemicals for fruit fly management (Chuman et al. 1987). Several traps, namely, jackson trap, liquid protein bait, dry synthetic lure/liquid protein, open bottom dry trap, yellow panel (Figure 9.3), cook and cunningham trap, champ trap, tephri trap and steiner trap, may be used.

9.3 WHITEFLIES

Whitefly (*Trialeuroides variabilis*), the minute winged pest of papaya is a sucking insect and it coats the leaves with honeydew which forms the basis for sooty mold development. Shaking young leaves will often reveal the presence of whiteflies. It is a major pest of the leaves of papaya trees. Damage to papaya caused by *T. variabilis* is similar to the damage commonly caused by whiteflies in other crops with heavy

FIGURE 9.3 (**See colour insert.**) Yellow panel trap for female fruit fly.

infestations; the leaves fall prematurely, fruit production is affected and their secretions promote the growth of sooty mold on foliage and fruits (Reiger 2006). Nymphs and adults suck the sap from undersurface of the leaves and yellowing may take place. Infested leaves are usually removed and appropriate pesticides applied to orchards.

Removal of host plants, installation of yellow sticky traps and spraying or dusting should begin when adults are noticed. At threshold level, the population of whiteflies can be controlled by foliar spray of imidacloprid (0.01%) or dimethoate (0.05%) or Metacystox (0.02%) at an interval of 10 days. Release of predators (*Coccinellid predator* and *Cryptolaemus* sp.) and parasitoids (*Encarsia haitierrsis* and *E. guadeloupae*) are very effective to control the severity of infestation (da Silva et al. 2007).

9.4 APHIDS

These are tiny lice-like insects with colour ranging from pink yellow to brown to black. Incidence of papaya ringspot virus (PRSV) coincides with the number of aphids trapped four week prior to infection suggesting a strong link between the aphid vectors and PRSV incidence (Mora-Aguilera et al. 1992; Kalleshwaraswamy et al. 2007). Since *A. gossypii* is numerically dominant vector in south India besides, being a more efficient vector capable of inoculating PRSV to multiple plants, it should be the target vector for control strategy. All the individual of *A. gossipii* or *A. craccivora* are potentially capable of PRSV transmission, though efficiency varied between individuals (Kalleshwaraswamy and Kumar 2007). PRSV is transmitted by several species of aphids in a non-persistent manner. This transmission is generally characterised by a short acquisition access period (AAP) of a few seconds to minutes, absence of a distinct latent period and a short inoculation access period (IAP). Thus, the virus is acquired during brief exploratory probes in the process of host finding and is transmitted to a healthy papaya plant within a few seconds to minutes. Aphids do not normally colonise papaya and natural transmission is reported to occur through transient aphid vectors.

9.4.1 Management

Seedlings raised under controlled conditions in net house significantly decrease PRSV incidence. Removal and destruction of the affected plants is the only control measure to reduce spread of the disease (Kumar et al. 2010). Aphids can be controlled by application of carbofuran (1.0 kg a.i./ha) in the nursery bed at the time of sowing seeds, followed by 2–3 foliar sprays of phosphomidon (0.05%) at an interval of 10 days, starting from 15 to 20 days after sowing.

9.5 SCALE INSECTS

More than five species of scale insects have been found on papayas, the most serious being the oriental scale, *Aonidiella orientalis*, which occurs on both fruit and stem. So far, it is confined to limited areas. Recently other scale insects, *Coccus hesperidum*, *Aonidiella comperei*, *Selenaspidus articulates*, *Aspidiotus destructor* and *Philaphedra tuberculosa* have been reported (Martins et al. 2004). Scale insects are not common on papaya trees but occasionally they build up in large numbers to cause serious damage or death to the trees. Scales do not resemble typical insect pests, so they often go unnoticed until a large population has developed and caused damage. In the virgin land, scale insect has been most troublesome, apart from rats and fruit-bats that attack ripe fruits.

Considerable damage may occur when large number of these insects feed upon plants already weakened by environmental stresses, such as drought or disease. Scales remove plant juices with piercing-sucking mouth parts. Some time, certain scale insects produce tremendous amounts of honeydew (an excess of liquid and sugar expelled from their bodies). This sticky material may serve as a growth medium for a sooty mold fungus. During large population build-up, some scales become so abundant that an infested plant is totally encrusted with them. Three types of damage to papaya plants are observed. Flower and leaf drop occur from severely infested young plants. Infestation on seedlings or on young plants near the apex observed, inducing distortion of apical leaves (Pena and Johnson 2006). Finally the females get attached to the fruit causing cosmetic damage that makes fruit unmarketable (Figure 9.4).

9.5.1 Management

The scale insect has at least nine different predators, among which the ladybird beetle (*C. montrouzieri*) is the most important. It is also attacked by fungus *Verticillium*

FIGURE 9.4 (**See colour insert.**) Papaya fruit infested with female scale insects. (Adapted from Pena, J. E. and F. A. Johnson, 2006, *Insect Management in Papaya*, The Institute of Food and Agricultural Sciences (IFAS), University of Florida, ENY-414:1–4.)

lecanii, which causes up to 90% mortality during summer. There are two small parasitic wasps, for example, *Coccophagous lycimnia* and *Trichomastus portoricensis* that periodically cause mortality to this pest (Copland and Ibrahim 1985). Foliar spray of malathion (1.5 mL/L of water) and lime sulphur (1.0 g/L water) has been reported very effective to control this pest (Pena and Johnson 2006).

9.6 MITES

Papaya foliage is attacked by a complex of phytophagus mites, among which the carmine spider mite (*Tetranychus cinnabarinus* Boisduval) is one of the key pests. This species is a worldwide polyphagous pest, mainly distributed in semitropical and tropical areas (Cheng et al. 2009). Tiny eight-legged yellow, dark green or reddish spiders inhabit the underside of foliage. Red spider mite (*Tetranychus seximaculatus*) sucks the sap from the leaves. In India, plant and fruit infestation by red spider mite has been a major problem. Mites prefer to suck from very young plant tissues. This pest feeds on the very young fruits and causes them to drop. They can also transmit viral diseases (Cheng et al. 2009).

9.6.1 Management

Heavily infested leaves should be pruned and buried or burnt. Planting of healthy and vigourous seedlings which are clean and without infestation or diseases should be used. Spraying of herbal insecticides and lime sulphur solution underneath the leaves will greatly help in cleaning the plants from these pests (Rex and Rivera 2005).

Control of mite pests on papayas depends mainly on chemical applications. However, intensive application of miticides, short life cycle and high reproductive rates of mites have led to the development of resistance in mites to many registered miticides (Goka 1998). The number of miticides that can be used is further limited because many miticides produce unacceptable phytotoxicity to papayas (Lo 2002). Therefore, it is necessary to search for alternative approaches for controlling papaya mite pests.

Removal of alternate host, ants and allowing natural predators and enemies to operate in the plantation is a welcome intervention. Three specialist predators are commonly found feeding on carmine spider mites in papaya, namely, the beetle *Stethorus siphonulus* (Kapur) and two predatory mites, *Phytoseiulus macropilis* Banks and *Phytoseiulus persimilis.* The introduced generalist predator, *Nesticodes rufipes,* a sit-and-wait tangle-web building spider, very abundant on papaya leaves, acts as the most effective predator because it can prey upon all members of the arthropod community. *M. basalis* can be successfully mass produced using a micro encapsulated artificial diet in a cost-effective manner (Lee 2003). Some insecticides, fungicides and acaricides have also been used (Lo 2002). *M. basalis* may be a compatible, viable candidate species for use in integrated pest management programs on papaya (Cheng et al. 2009).

9.7 WEBWORM

The papaya webworm (*Homolapalpia dalera*) also known as fruit cluster worm is mainly a widespread pest of the developing fruit peel and papaya stem. It is usually found near the stem amongst the flowers and fruits. The webworm causes damage by eating the stems and the fruits of the papaya plant (Morton 1987). The holes created by this worm allow the anthracnose fungus to enter the affected parts of the plant, causing further damage and sunken spots on the fruit.

9.7.1 Management

The presence of webworm is evident by the presence of webs and damage can be seen near the main stem, the region of attachment of each of the fruits produced by the plant. Control includes hand removal and hosing off the plant with a strong stream of water from a garden hose (Crane 2005). Damage can be prevented if 2–3 foliar sprays of dimethoate (1.0 mL/L of water) at the beginning of fruit set, or at least at the first sign of webs are undertaken. The best method of treatment to prevent webworm is by spraying the plant with malathion when the plant is about to start producing fruits. Successful reduction of webworm can result from the use of malathion (at 1.5 mL/L of water) or *Bacillus thruingiensis* (Bt).

9.8 NEMATODES

Papaya suffers from many pests including plant parasitic nematodes and often crop losses may be substantial. Papaya trees are susceptible to several nematodes, but only two genera appear to cause economic damage in papaya cultivation. They are reniform (*Rotylenchulus reniformis*) and root-knot (*Meloidogyne incognita*), both of which have a worldwide distribution in papaya plantations (Perera et al. 2008).

Root-knot nematodes are one of the most important nematodes associated with papaya (Coveness 1967).

9.8.1 Root-Knot Nematode

A root-knot nematode infestation under irrigated conditions primarily caused by *M. javanica* and *M. incognita* has been reported from many countries (Morton 1987). These nematodes attack the root systems of the plant and impair water and nutrient uptake. Heavy nematode infestations can cause wilting, stunting, decreased plant vigor, reduced yields and shortening of the productive life of a papaya tree. Young plants of papaya are severely damaged as compared to the mature trees. The larvae of these nematodes can travel short distances in soil, finding and attacking papaya roots only, near the tips. When female larvae feed near the water conducting core of the roots, the plant cells increase in number and size until readily visible swellings, called galls or 'knots' are formed. Small to large swellings (galls), which are produced as a result of the feeding process of the nematode, interfere with the proper functioning of the roots. Severe attack of this nature causes retarded root growth and a subsequent reduced root system for the plant (Figure 9.5). Secondary symptoms which may be seen in the above ground portion of the tree are like those associated with trees suffering from malnutrition or water stress. Leaves of papayas that are affected appear pale green or slightly yellow and may fall prematurely. Infected plants are sensitive to slight moisture stresses and wilt more readily than the non-infected plants. Fruits produced are smaller than normal, may be slightly insipid and more likely to have an off-flavour. These nematodes are principally spread through cultivation and surface runoff or irrigation.

9.8.2 Management

Summer ploughing and exposing the soil to sunlight for one or two months during April–May reduce the nematode and pathogen load in soil. Crop rotation with onions, baby corn, sweet corn, maize, millet, sorghum, sesame, cassava and sudan grass reduces nematode incidence for the next crop (Dobson et al. 2002). Mixed cropping with marigold or Indian mustard as trap crops, can also minimise root-knot nematode damage. Purchase of nematode-free papaya seedlings for transplanting in orchards, application of farmyard manure (FYM), neem cake, pressmud and carbofuran 3G at 8.0, 1.0, 3.0 and 10.0 g, respectively, per plant at the time of planting in the pits are very effective (Singh et al. 2010a,b). Application of carbofuran 3G at 40.0 g/tree before flowering is also an effective method of its management. Use of resistant cultivars, namely, Solo, Pusa Majesty, Washington and Coorg Honeydew against *R. reniformis* and CO-2 and CO-3 against *M. incognita* (Ram 2005) are recommended in susceptible areas.

Use of chemicals as preplanting soil fumigation is effective (Ahmad and Sultana 1981) but recent concerns about their indiscriminate use necessitates a careful approach. These were also known to affect beneficial microflora and, thus, disturb the ecological balance. Control of nematodes using plant-based material is suggested which increases soil fertility (Muller and Gooch 1982) as well as causes considerable

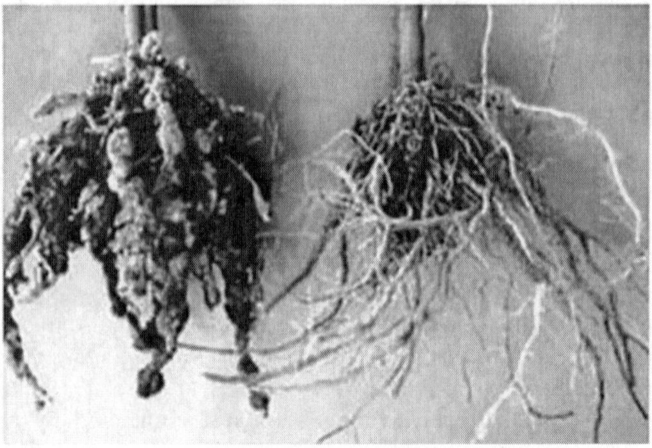

FIGURE 9.5 Root-knot nematode infections on roots of papaya. (From Dobson, H. et al., 2002, *Integrated Vegetable Pest Management*, Natural Resources Institute, University of Greenwich, UK.)

deleterious effect on plant nematode populations. Control that is economically feasible has been obtained by using 1,2-dibromo-3-chloropropane (Nemagon) at 20.0–25.0 kg per acre, a mixture of dichloropropane and dichloropropene (D–D) at 80.0–85.0 kg per acre.

9.9 EARTHWORMS

The immature earthworms (*Megascolex insignis*) feed on rotting tissues of papaya plants and they hasten the demise of plants affected with stem rot from *Pythium aphanidermatum* and may act as vectors for this fungus. Some time, the nurseries of papaya are also disturbed and damaged through gallery formation by earthworm.

Application of carbofuran 3G (10.0 g/m^2 nursery area) during bed preparation to overcome the damage through gallery formation by earthworm was very effective. Application of carbofuran 3G at (40.0 g/tree) at pit filling and before flowering is also effective method for stem rot vector management (Ram 2005).

9.10 ANTS

Ants rarely feed directly on plants, but can cause damage in other ways. There are many species and they lead complicated lives with fascinating social structures. Papaya is infested with insects that excrete honeydew; ants can often be seen collecting it and protecting the insects that produce this sugary substance. By eating the honeydew, ants actually help plants. Honeydew, while not damaging in itself, supports the growth of sooty moulds. These moulds have dark spores which turn the leaves black. The shading of the leaves by these spores reduces photosynthesis. Ants can be small to large, yellow, red brown or black, winged or wingless live in colonies. They may also act as vectors for mealy bugs, fungus, etc. Their infestation can be kept to minimum by regular hand weeding and cultivation at base of the plant to disturb foraging ants, their nests and maintaining a weed free area at the base of each plant, the diameter of which is the same as the plant canopy.

9.11 BIRDS AND ANIMALS

Papaya saplings are damaged by snails (Figure 9.6a), birds, squirrels, monkeys, elephants, etc. Crows become a great nuisance during winter months when they gather in order to stay warmer. Crows mostly damage the ripe fruits by eating pulp and also seeds of papaya (Figure 9.6b, c). At early stage of fruit development, they damage the fruit surface by repeated injuries with their beaks (causing cosmetic damage that makes fruit unmarketable).

9.11.1 Management

Covering fruits with thatch, gunny bags or festoon of thorny sticks or nylon nets and scaring facilitates the management of vertebrate pests. Always remove food sources including rotting fruits, etc., from and around the papaya fields. Never allow animals of any sort of easy access to food. Trash cans and bird feeders should all be covered and removed. Do not expect animals to leave the fields, if there are plenty of food and water sources. In addition to recorded distress or alarm calls, frightening devices include gas-operated exploders, battery-operated alarms, chemical frightening agents, lights (for roosting sites at night), bright objects, clapper devices and various other noise makers like beating of tin sheets or barrels with clubs can help in scaring birds. Spraying birds as they land, with water from a hose or from sprinklers mounted in the roost trees, has helped in some situations. Hanging Mylar tape in roost trees may be helpful in urban areas. A combination of several scare techniques used together work better than a single technique used alone. Vary the location,

FIGURE 9.6 **(See colour insert.)** Snails (a) and crows (b, c) eating seeds and fruits during winter in a papaya orchard.

intensity and types of scare devices to improve their effectiveness. Supplement frightening techniques with shotguns, where permitted, to improve their effectiveness in dispersing crows. Ultrasonic (high frequency, above 20 kHz) sounds are not effective in frightening crows and most other birds because, like humans, they do not hear these sounds. Cassette tapes used with pyrotechnics can be very effective. Your neighbours may not like it as it will make the crows move elsewhere. Tapes can also be used to attract crows for hunting purposes. Again this is a short-term solution testing up to a year. Repellents are also an option. Essentially, it is a chemical that is fogged and sent up into the trees. One problem with repellents is that we may not be able to use them in situations where fruits are grown for eating or marketing. Shooting the crows is safe and legal. Trapping can be an effective option (Yahner and Wright 1985).

9.11.1.1 Phytotoxicity

Papaya-growing regions have to combat a number of major and minor pests and diseases, and the extent of damage depends upon agroecological regions. The common strategy for pest management, sometimes, is prevention through the application of pesticides at regular intervals. Proper timing, deposition and coverage of the pesticides' application are critical for the effective control. Reproductive precocity, high photosynthetic rates of short-lived leaves, fast growth and high reproductive

FIGURE 9.7 **(See colour insert.)** Symptoms of phytotoxicity.

output, production of many seeds and low construction cost of hollow stems, petioles and fruits production due to faster absorption of water, nutrients and chemicals as applied and available in soil make it very sensitive against inorganic pesticides and nutrients. Papaya plants are very sensitive to pesticide(s). Particularly, two pesticides diamethoate for aphid control and omite for management of mites have shown phototoxic effect on young papaya plants. High dose and repeated use of diamethoate (≥2.0 mL/L of water) cause toxicity on newly emerged leaves showing light green colour of leaf veins (Figure 9.7a and b), but the damaged leaves of papaya recover within 15–20 days after application. Therefore, we should use recommended pesticides only. Sometimes, farmers and extension workers used to apply chlorpyriphos (at 1.5–2.0 mL/L of water) in papaya orchards (Figure 9.7d) for mealy bug control in papaya and other fruit crops. Papaya plants are also very sensitive to soil application of boron, especially in sandy to sandy loam soils. Soil properties such as texture, temperature, moisture, microbial activity and pH may also influence phytotoxicity. At higher pH, soils are less binding hence may enhance phytotoxicity, while high microbial activity may reduce it. Basal application of borax at ≥6.0 g/plant causes burning of leaf tip and margins (Figure 9.7c). Seed germination of papaya was severely affected by seed coating and treatment by propiconazole at 1.5–2.0 mL/kg of seeds before sowing in nursery (Figure 9.7e). Phytotoxic properties of pesticides are usually associated with specific formulations or specific plants rather than group of pesticides or plants.

Least toxic pesticides (natural or organic in nature) are not necessarily harmful to plants or environment. Many are quite safe to use. Calibration of the equipment and chemical quantity prior to each application is important in order to avoid heavy stresses, fruit damages and burning of leaves. A pesticide label may indicate whether the pesticide is phytotoxic and may list plants or varieties that are sensitive. Reading and following label directions, especially the correct rates and timing, and being aware of potential weather effects, avoiding application of pesticides when drift is likely to happen, waiting for the correct planting times. If unsure, it is always advisable to conduct a simple field bioassay by treating only a few plants, before treating the whole block for phytotoxic effects, especially when growing new crops and cultivars.

10 Diseases

The diseases of papaya include those caused by viruses and fungi. Papaya growing areas have to combat a number of major and minor diseases, and the extent of damage depends upon agroecological regions. Fungal diseases are a major problem in papaya production. The common strategy for fungal disease management is prevention through the application of fungicides at regular intervals. Proper timing, deposition and coverage of the fungicide application are critical for the effective control. Once the disease is established, the fungicides have little impact on control. Viral diseases have been recognised as the greatest threat to the papaya industry. It is thought that at least two viral diseases are involved in papaya and it has been suggested that the diseases are spread in part by the tapping of green fruits for their latex (the source of papain). Most serious of all is the papaya ringspot virus (PRSV) which affects both plant and fruits.

10.1 PAPAYA RINGSPOT VIRUS

The term papaya ringspot (PRS) was first coined by Jensen in 1949 to describe a papaya disease in Hawaii. The inevitable entry of *Papaya ringspot virus* (PRSV) into the Puna district on Hawaii Island was discovered during the first week of May in a papaya field in Pahoa, 1–3 miles from the major papaya growing areas in Puna (Hawaii). Diseases such as papaya mosaic (caused by *Papaya mosaic virus* (PapMV)) and watermelon mosaic (caused by *Watermelon mosaic virus*-1) were shown recently to be caused by PRSV. The PRSV causes a major disease of papaya and cucurbits and is found in all areas of the world where papaya and cucurbits are cultivated. The primary host range of PRSV is limited to papaya (Caricaceae) and cucurbits (Cucurbitaceae), with *Chenopium amaranticolor* and *C. quinoa* (Chenopodiaceae) serving as local lesion hosts. The virus is grouped into the papaya infecting biotype (PRSV-p) which affects papaya and/or cucurbits, and the cucurbit infecting biotype (PRSV-w) which affects cucurbits only. PRSV belongs to family Potyviridae, genus potyvirus and is an aphid-transmissible RNA virus that commonly infects papaya, causing serious disease and economic loss. Virions of PRSV are rod-shaped and flexuous measuring 760–800 × 12 nm with a monopartite single-stranded positive sense RNA as its genome. Like other potyviruses, PRSV is transmitted in a non-persistent manner by several species of aphids. Genetically engineered papaya has been used to successfully control the disease caused by PRSV in Hawaii. The virus, PRSV, is transmitted non-persistently by aphid vectors and does not multiply in the vector. The disease cycle can start with aphids feeding on infected papaya for as little as 15 s and subsequently feeding on a healthy papaya. There is no incubation period. The virus does not persist in the vector so transmission to another plant has to occur rather rapidly (Gonsalves et al. 2010).

10.1.1 Symptoms

The earliest symptoms on papaya are yellowing and vein-clearing of the young leaves. In papaya, leaves develop prominent mosaic and chlorosis on the leaf lamina, and water-soaked oily streaks on the petioles and upper part of the trunk. Severe symptoms often include a distortion of young leaves which also result in the development of a shoestring appearance that resembles mite damage (Figure 10.1). Trees that are infected at a young stage remain stunted and will not produce an economical crop. Fruits from infected trees may have bumps similar to that observed on fruit of plants with boron deficiency and often have 'ring spots', which is the basis for the disease's common name. A severe PRSV isolate from Taiwan is also known to induce systemic necrosis and wilting along with mosaic and chlorosis (Gonsalves 1998). The disease derives its name from the striking symptoms that develop on fruit. These consist of concentric rings and spots or C-shaped markings, a darker green than the background-green fruit colour. Symptoms persist on the ripe fruit as darker orange-brown rings. Vigour of trees and fruit set is usually reduced depending on the age of the plant when infected. Fruit quality, particularly flavour, is also adversely affected. If affected plants are not removed, the condition spreads throughout the plantation. Fruits borne 2 or 3 months after the first symptoms have a disagreeable or bitter flavour in taste.

Papaya mosaic symptoms were first reported in 1962 together with another virus that became known as PRSV, a potyvirus. The two viruses were differentiated in 1965 and 1967 using particle lengths, serology, host range, inclusion bodies and aphid transmissibility. PapMV causes mild mosaic symptoms on leaves and stunting of the plant. Presently, it is also known as PRSV. Recently, both the viruses were reported at molecular level to be the same as potyvirus.

A mosaic symptom is more serious on young plants. The disease symptoms appear on the top young leaves of the plants. The leaves are reduced in size and show blister like patches of dark-green tissue, alternating with yellowish-green lamina (Figure 10.2). The leaf petiole is reduced in length and the top leaves assume an upright position. The infected plants show a marked reduction in growth (Noa-Carrazana et al. 2006). The fruits borne on diseased plants develop water soaked lesions with a central solid spot. Infected fruits are elongated and reduced in size.

FIGURE 10.1 Symptoms of PRSV on papaya leaves and fruits.

FIGURE 10.2 **(See colour insert.)** Mosaic symptoms on papaya plant.

10.1.2 Management

Papaya is not a preferred host of the PRSV vectors and aphids. Disease management by controlling the vector within the papaya orchard is very difficult as aphids visit and probe papayas but do not colonise papaya orchards. Papaya plants infected with PRSV must be destroyed in order to minimise spread of the virus. Virus problems can be avoided by planting genetically resistant cultivars. PRSV is not transmitted via seeds, but it can be spread to areas where it is not present by transporting infected seedlings. To avoid introductions of the virus, do not transport diseased papaya seedlings to different areas. In areas where PRSV is already present, raise seedlings in nurseries close to the planting site to maximise the possibility of spreading the viruses further. Early detection of infected plants and prompt removal can check the further spread of the disease. The various options for managing viral diseases are vector control, planting in areas where there is no virus, roguing, netting, cross protection and resistance (Tripathi et al. 2008; Gonsalves et al. 2010). Aphids can be controlled by application of Carbofuran (1.0 kg a.i./ha) at the time of sowing seeds, followed by 2–3 foliar sprays of dimethoate or phosphamidon (0.05%) at an interval of 10 days starting from 15–20 days after sowing (NHB 2002). As mentioned above, the economically relevant host range of PRSV includes papaya and cucurbits, two crops that are very different in growth characteristics. PRS-resistant genetically engineered papaya lines have been designed like, 55-1 and 63-2 using virus coat protein. Papaya ringspot potyvirus gene (CaMV 35S uidA) leader sequence was identified at Hawaii. These transgenics were produced by biolistic (particle bombardment) transformation of embryogenic cultures of the papaya cultivar, 'Sunset', with DNA-coated tungsten particles. Papaya lines, 55-1 and 63-1, were used to create the cultivars, SunUp and

Rainbow. SunUp is a transgenic version of its red-fleshed, progenitor cultivar, Sunset and is homozygous for the coat protein gene. Rainbow is an F_1 hybrid created by crossing transgenic SunUp with non-transgenic cultivar, Kapoho.

Cross protection is the phenomenon whereby plants that are systemically infected with a mild strain of virus are protected against the effects of infection by a more virulent related strain. Cross protection was practiced quite extensively in Taiwan and to a very limited extent in Hawaii in the 1980s. At both locations, the same mild strain was used, which was developed by selecting a mild mutant from nitrous acid treatment of tissue extracts of leaves infected with a severe strain of PRSV in Hawaii. This mutant strain caused mild symptoms on both papaya and cucurbits and gave good cross protection against severe strains of PRSV in Hawaii. However, the first attempts to use this mild strain for large scale protection was initiated in Taiwan, where PRSV was widespread and was causing severe damage. Even in Hawaii, however, the practice was not fully commercialised because the mild strain caused quite severe symptoms on Sunrise cultivar and because the logistics for raising the inoculum, uniformly infecting plants with the mild strain and delivery of the mild strain-infected plants to the farmers did not work out economically (Tripathi et al. 2008; Gonsalves et al. 2010). Field sanitation such as removal and destruction of affected plants reduce the spread of the disease (Bau et al. 2008).

10.2 LEAF CURL

The disease is transmitted by the vector, whitefly (*Bemisia tabaci*). The virus is not seedborne. The disease was characterised by severe curling, crinkling and deformation of the leaves (Figure 10.3). Mostly the young leaves are affected. Apart from curling, the leaves also exhibit vein clearing and thickening of the veins. Sometimes

FIGURE 10.3 Leaf curl in plant cv., Pusa Dwarf.

the petioles are twisted. In severe cases, complete defoliation of the affected plant is observed. The affected plants show a stunted growth with reduced fruit yield and quality (Ram 2005).

10.2.1 Management

Practically, there is no control measure for this virus. Removal (rouging) and destruction of the affected plants is the only control measure to reduce the spread of the disease. It can also be minimised by planting the crop after rainy season (Ram 1984a, b) . Checking the population of whitefly can also reduce the infection severity. Application of dimethoate (0.05%) or metasystox (0.02%) at an interval of 10 days effectively controls the whitefly population (NHB 2002). Addition of heavy doses of organic matter results in lesser disease incidence as compared to the application of chemical fertilisers.

10.3 PAPAYA APICAL NECROSIS

Papaya apical necrosis is a relatively new virus. Symptoms include a drooping and downward cupping of the leaves, reduced leaf size and browning of the leaf margins. At present, there is no control for this disease.

10.4 COLLAR ROT OR FOOT ROT

This disease is caused by soil-borne fungi namely *Pythium aphanidermatum* and *Phytophthora palmivora* (Guadalupe 1981). The oospores are formed on the papaya residue in soil. The pathogen can survive on dead organic matters as saprophyte and causes infection when suitable host is grown in such soil. The secondary spread takes place by zoospores. This disease is widely spread in papaya plantations of India, Sri Lanka, Hawaii and South Africa. Under favourable conditions of high rain fall and high temperature, the whole plantation is wiped out within one season. The incidence of this disease would be severe during rainy season and commonly found in poor-drained soils (Erwin and Ribeiro 1996). It is characterised by the appearance of water-soaked patches on the stem near the ground level. These patches enlarge rapidly and girdle the stem, causing rotting of the tissues, which then turn dark brown or black. Such affected plants cannot withstand strong wind and topple over and die. If the disease attack is mild, only one side of the stem rots and the plants remain stunted. Fruits are shriveled, malformed and gradually the plant dies. Trunk rot (*P. ultimum*) is damaging to papaya in USA, Africa and India. It is also known as collar rot in 8–10 months old seedlings, evidenced by stunting, leaf-yellowing, shedding and total loss of roots (Koffi et al. 2010).

10.4.1 Management

Papaya is, generally, not recommended on land previously planted and this is due to soil infestation by *P. aphanidermatum* and *P. palmivora*. Plant refuses from previous plantings should never be incorporated into the soil. In the case of new plantings, preventing water logging of the soil may control the disease. Soil fumigation is necessary before replanting papaya in the same field. Application of *Trichoderma viride* (15.0 g/plant) mixed in well-decomposed farmyard manure (FYM) should be applied around the root zone of the plants at the time of planting. The soil should be drenched with 2–3 L of copper oxychloride (2.5 g/L of water). The application should be carried out regularly at 15 days interval from the time of planting. During fruit formation, the plant should be sprayed with the same solution at the same time interval (Erwin and Ribeiro 1996). Alternately, mancozeb (2.5 g/L of water) may also be applied. In the case of coller rot/stem rot attack in existing crops, the rotted portion of the plant should be scraped and copper oxychloride or bordeaux paste should be applied (Ram 2005). The paste can be prepared by dissolving 1.0 kg of copper sulphate and 1.0 kg lime separately in 10 L of water each. The two solutions should be mixed and shaken to form a paste. We have observed that root drenching with ridomil MZ @ 2.5 g/L of water is also very effective. *Phytophthora* is effectively managed by application of metalaxyl M + phosphonate (Erwin and Ribeiro 1996). This fungicide is a xylem-translocated compound with an upward movement in plants in the transpiration stream (Guest et al. 1995). Thus, metalaxyl and related acylanilide compounds have no effect on root diseases if applied as a foliar spray because they are not transported to the roots. Phosphonates will not eradicate the pathogen or eliminate disease, but remain an excellent, cost-effective option for control of phytophthora diseases (Drenth and Guest 2004).

10.5 BACTERIAL WILT

This disease, also known as wilting, chlorosis and necrosis, is caused by the bacteria *Pseudomonas solanacearum* and was discovered by E.F. Smith (Seshadri et al. 1977; Nelson 2012a, b; APS 2014). The incidence of this disease can be severe during the rainy season and very commonly found in papaya plantation raised in poor-drained soils. Wilt-like symptoms are very severe in India, especially in Bihar, as growers wish to abandon their plantations. It is characterised by symptoms *viz*., yellowing of upper leaves (Figure 10.4a), followed by death of leaves and subsequent defoliation (Nelson 2012a, b). The root system was extensively rotted starting first from tertiary to secondary roots and finally to tap roots (Figure 10.4b). Fruits get shriveled, drop their unripe fruits gradually, and plant dies within 10–15 days (Figure 10.4c). Further research for this disease is required.

10.5.1 Management

Generally, papaya is recommended three-year crop rotation and this is to avoid soil infestation on land previously planted. The crop should be irrigated by adopting the ring method of irrigation so that the water does not come in direct contact with the stem. In the case of new plantings, preventing water logging of the soil may control the disease. Plant refuses/residue from previous plantings should never be incorporated into the soil. It has been observed that the root drenching with combination of copper oxychloride (2.0 g/L of water) + streptocycline (0.03 g/L of water) is effective to recover the plants (Figure 10.5).

FIGURE 10.4 (**See colour insert.**) (a) Root tip, (b) root rot initial stage and (c) later stage.

FIGURE 10.5 Recovery of plant after treatment.

10.6 POWDERY MILDEW

This disease is caused by *Oidium caricae* (imperfect state, *Erysiphe cruciferarum*) often affects papaya plants and fruits serious in the humid tropics. Sometimes, other species of fungus, *Golovinomyces cichoracearum*, *Oidiopsis sicula, O. caricae, O. caricae-papayae, O. caricicola, O. indicum, O. papayae, Ovulariopsis papayae, P. caricae-papayae, P. macularis, P. xanthii* and *Streptopodium caricae* were also reported to infect papaya from Taiwan (Tsay et al. 2011). The development of powdery mildew in papaya is promoted at high humidity (80%–85%) and a temperature range of 24–26°C. Infection is first apparent on undersurfaces of leaves as small, slightly darkened areas, which later become white powdery spots (Rawal 2010). These spots enlarge and cover the entire leaf area. Severely infected leaves may become chlorotic and distorted before falling. Fungus may also attack the stem and petioles of young plants under reduced light. Affected fruits are small in size and malformed. Fungus takes nutrients from the cells of the leaf surface by haustoria and produces mass of spores which are carried out by wind from diseased to healthy plants.

10.6.1 Management

In the initial stage of disease, dusting or spraying of sulphur (2.0 g/L of water) at 15 days interval helps to control the disease (Rawal 2010). It may also be controlled by spraying of Karathen @1.0 mL/L of water (Ram 2005).

10.7 DAMPING-OFF

Damping-off can occur at the pre-emergence stage, in which case seedlings do not emerge, or post-emergence where seedlings will fall over and die from a constriction on the stem (often brown) at soil level (Figure 10.6). The tissues of papaya seedling, arises from soil line, become water-soaked and collapse due to the growth of fungus. The disease is caused by a complex of fungi viz., *P. aphanidermatum, P. palmivora, Rhizoctonia solani* and *Fusarium* sp. (DAIS 2009). Commonly, this disease is caused by *Rhizoctonia*. Lesions are seen on the stem at or just above soil level. This is a soil-borne disease. The stem becomes watery and shrinks, followed by death of the plant. This disease is aggravated by excessive water or fertiliser, heavy soils, overcrowding in seedling beds, wet weather and high-humidity conditions (Rawal 2010).

FIGURE 10.6 Post-emergence stage of damping-off in nursery.

10.7.1 Management

Nursery bed should be treated with formaldehyde (10%) before sowing. Well-drained soil should be used for planting and the crop should not be excessively irrigated. Before sowing, the seeds should be treated with bio-control agent, that is, *T. viride* (5.0 g/kg of seed) or captan (3.0 g/kg of seed) to protect the newly emerging seedlings. Drenching of nursery with a combination of carbandazim (12%) and mancozeb (64% W.P.) @ 2.0 g/L of water effectively controls this disease 12 DAS of seeds. This can also be controlled by spraying bordeaux mixture (5:5:50) 12 DAS in nursery (Ram 2005).

10.8 BUD AND FRUIT STALK ROT OF PAPAYA

This is a fungal disease (*Fusarium solani*) which affects papaya buds severely (Ram 1984). The infection affects stalk and newly born fruits drop down. The disease first manifests as pale yellow discolouration near the base of stalk, which later on spreads to the whole stalk. Afterwards, some stalks turn dark brown to black. The flower buds in early stages are normally infected. The corolla and rudimentary calyx of such a flower turns yellow, dries and finally drops. The infection spreads to the ovary, which gets shriveled and mummified and ultimately drops. The young papaya fruits are also affected (Figure 10.7a). The infection starts as water-soaked lesions appear on the skin with appearance of fructification. Gradually, the infected portion turns brown to black and sunken. The fruits decay and fall. The disease causes severe losses during rainy season. Plants remain senile during rainy months due to infection of this disease (Figure 10.7b).

10.8.1 Management

Prophylactic spray of Bordeaux mixture (5%) or copper oxychloride (3.0 g/L of water) before onset of rainfall and repetition at 15 days interval effectively control this disease (Ram 1984).

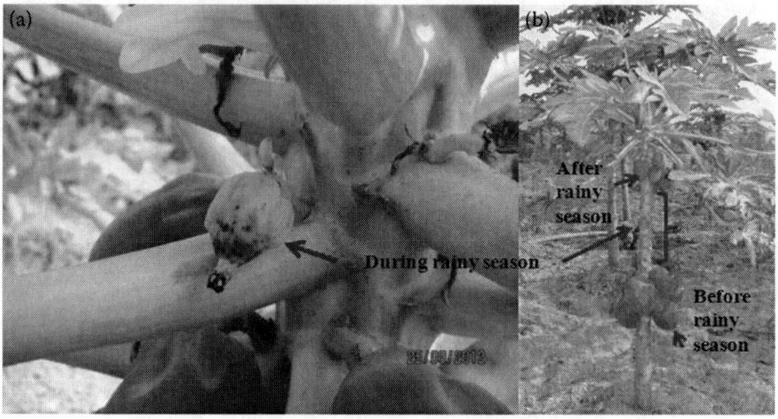

FIGURE 10.7 Bud rot in papaya: (a) infected bud and (b) bud drop during rainy season.

10.9 ANTHRACNOSE

Anthracnose is the most serious disease affecting the ripened fruit caused by *Colletotrichum gloeosporioides* Penz., in which light yellow patch appears on the south west side exposed to sun in India. It is an important papaya disease in Hawaii, Mexico and India (Rawal 2010). This fungus attacks not only the fruit on which it causes the most damage, but also the petioles of the low, older leaves that begin to turn yellow. The disease prominently appears on green immature fruits (Figure 10.8a).

10.9.1 Symptoms

The disease symptoms are in the form of brown to black depressed spots on the fruits. The initial symptoms are water-soaked, sunken spots on the fruit. The patches slowly soften, turn brown and extend to half fruit. The centers of these spots later turn black and then pink when fungus produces spores (Figure 10.8b). The flesh beneath the spots becomes soft and watery, which spreads to the entire fruit (Hasan et al. 2012). Small, irregular-shaped, water-soaked spots on leaves may also be seen. These spots eventually turn brown. On the fruits, symptoms appear only upon ripening and may not be apparent at the time of harvest. Brown sunken spots develop on the fruit surface, which later on enlarge to form water-soaked lesions. The flesh beneath the affected portion becomes soft and begins to rot (NHB 2002). A disease resembling anthracnose but which attacks papayas when just beginning to ripen, was reported from the Philippines in 1974 and the causal agent was identified as *Fusarium solani*. A strain of this fungus produces 'chocolate spot' (small, angular and superficial lesions) on fruits. The fungus causing anthracnose on fruit also attacks the petioles of lower leaves as they begin to die and shed from the plant.

FIGURE 10.8 **(See colour insert.)** Fruits damaged by anthracnose disease: (a) unripe and (b) ripe stage.

Infection on these petioles is important since they may act as a source of inoculum for infection of fruit. The trunk also gets scorched on the side exposed to sun and develops sunken area on its bark which gets dry and fibrous, causing the plants to collapse during the summer season.

10.9.2 Management

The affected fruits should be removed and destroyed properly. The fruits should be harvested as soon as they mature. Since the disease is associated with scorching sun, shading the fruits and trunk is a preventing measure. Closer planting and frequent irrigations may reduce this malady. Ammonium carbonate (3.0%) incorporated into the wax formulation effectively reduced anthracnose incidence by 70% in naturally infected papaya and extended the storage life by maintaining the firmness, color and overall quality of the fruit in low temperature storage (13.5°C) and 95% R.H. for 21 days, followed by 2 days, under marketing conditions (Shiva Kumar et al. 2002). It can also be controlled by propiconazol (Sepiah 1993), double dip hot water treatment and hot water dip treatment at 43–49°C for 20 min with combination of fungicide (Couey et al. 1984). However, hot drip water treatment also affects the ripening process (Paull 1990). Sodium bicarbonate (2%) significantly reduced the disease incidence and severity up to 60% during storage in papaya (Hasan et al. 2012). In the initial stage of this disease, it can also be controlled by spraying with mancozeb (1.0 g/L of water) or carbendazim (1.0 g/L of water) or thiophanate methyl (1.0 g/L of water) at 15 days interval (NHB 2002).

10.10 LEAF BLIGHT

The disease is caused by *Corynespora cassiicola* and severely damages the leaves. The disease first appears as small, discoloured lesions, which are irregularly scattered on the leaves. These spots become irregular in shape, then increase in size and appear brown to grey in colour. A light yellow zone surrounds the spots. Several lesions coalesce to cover large areas of the leaf and in severe infections; the whole leaf dies (Figure 10.9). A considerable reduction in the yield is observed. It can be controlled by spraying of mancozeb (2.0 g/L of water) starting from the appearance of the disease symptoms.

10.11 PHYTOPHTHORA BLIGHT

Phytophthora blight has also been called soft foot rot, stem canker, soft fruit rot and root rot. The pathogen, *P. palmivora* was named in 1919 by E.J. Butler. It was once classified as a fungus, but now it is regarded as a pseudo fungus in the stramenopiles. It is caused by a fungus (*Phytophthora parasitica*) and major disease in high-rainfall areas and wet weather. Young fruits are affected by this disease with water-soaked lesions exuding milky latex. Fruit rot initially appears as small, circular water-soaked lesions about 5–10 mm in diameter. Large lesions, often forming first where the fruit contacts the stem of the plant, are covered with whitish mycelium and masses of *Phytophthora sporangia* (Erwin and Ribeiro 1996). Fruits can rot, turn soft and fall prematurely and mummify. The top portion of the fruit-bearing

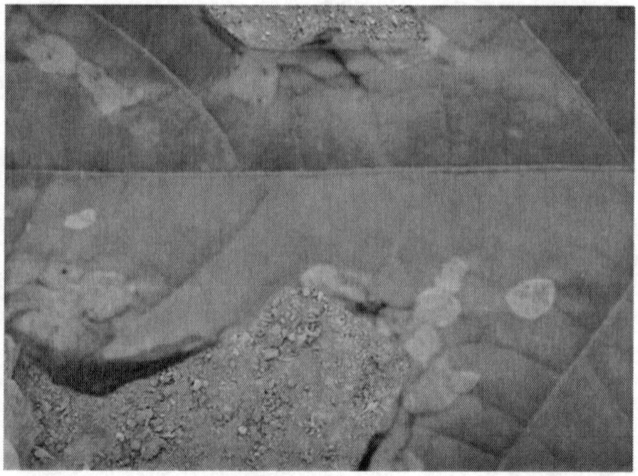

FIGURE 10.9 Leaf damaged by leaf blight disease.

region of the stem is susceptible to infection during rainy periods and causes stem canker like appearance. The infected plant may become more susceptible to wind damage (Hunter and Kunimoto 1974). Older portions of stems are susceptible when wet after extended rainfall/injury. Foliage on affected stems may collapse. Lateral roots of young plants are most susceptible in poorly drained soils. Roots may become dark and rotten, causing stunting of plant growth and leaves turn yellow and collapse. Severely infected plants may die. Plants with a heavy load of fruit may topple (Figure 10.10). Papaya plants with rotten roots become susceptible to drought stress.

10.11.1 Management

Select a well-drained low-rainfall site for cultivation of papaya. Orchard should be planted in well-drained soil with improved drainage in problematic soils is important

FIGURE 10.10 Papaya field damaged by *Phytophthora* blight.

to prevent damping off of seedlings. Pick up, remove and destroy fallen fruits, especially those with diseased symptoms. Intercrop papaya with non-susceptible host plants and avoid successive crops in the same field (Ko 1982). Seedbeds in nurseries should be fumigated by formaldehyde (10.0 mL/L water) prior to planting and avoid mechanical injury during cultivation practices (Nelson 2008).

10.11.2 Chemical Control

In high-rainfall and humid areas, preventive sprays of mancozeb and copper oxychloride @ 2.5 g/L water should be used at about every 2 week interval. Once *Phytophthora* blight appears in a field, the disease can become a major concern due to its ability to spread among plants and destroy fruits rapidly during windy, rainy periods. In that case, curative, systemic metalaxyl fungicides (Ridomil MZ) @ 2.0 g/L water would be very effective (Nishijima 1999). Under high-rainfall conditions, high-volume sprays are required at 2 week intervals. Under drier conditions, low-volume sprays at 4 week intervals are adequate to protect the exposed fruit surfaces. The use of a surfactant is important to ensure good distribution and adherence of the fungicide spray.

10.12 FRECKLES

Papayas are frequently blemished by a condition called ‘freckles’ of unknown origin and mysterious hard lumps of varying sizes and forms may be found in ripe fruits. No freckles were detected on young fruit; the skin area got affected and freckle diameter (FD) increased during the last phase of fruit growth as the fruit approached maturity. More freckles were seen on the exposed side of the fruit away from the stem (Eloisa et al. 1994). These spots are apparently associated with stomata. Star spots (greyish-white, star-shaped superficial markings) appear on immature fruits after exposure to cold winter winds. As the fruit matures, these spots may vary from pinpoint to >5.0 cm in size with a reticulate pattern. Frequently, a large irregular-shaped, water-soaked, or greasy spot area may surround several smaller dark spots. The spots are essentially brown in color on the green or maturing fruit. In the larger spots, the central portion may take on a greyish cast (Hine et al. 1965). Freckles are more prevalent on the exposed surface of the fruit as it hangs on the tree. In Uttar Pradesh, an alga, *Cephaleuros mycoidea*, often disfigures the fruit surface. Repeated isolations have failed to consistently yield an organism of a possible parasitic nature.

11 Physiological Disorders

Some environmental and nutritional stresses cause several physiological disorders, namely, bumpy, carpelloid, lumpy and deformed fruits, vivipary and white seeds. These are emerging physiological anomalies of developing papaya fruits and seeds in seed production.

11.1 BUMPY FRUIT

Boron has significant influence over production of uniform and healthy fruits, besides this, it also increases the production of fruits. Bumpy fruit of papaya is associated with boron deficiency. It occurs in papaya-growing areas of the world. The germplasm line, Pune selection-3 was the most susceptible under Bihar conditions against this disorder (Table 11.1). It may be due to comparatively higher boron requirement of this genotype.

11.1.1 Symptoms

First symptoms expressed are moderate chlorosis of mature in ripe leaves, followed by deformation of leaves, becoming fragile and curved towards back side and bleeding latex on leaves, stem and petioles. Death of apical sprout allows a lateral blooming. Fruit deformity first starts in young fruits, but symptoms become more severe on fruits close to physiological ripening (6–9 months at onset of fruit harvest). The bumpiness begins on the fruit epidermis due to boron deficiency, stopping the fruit growth. In addition to this, infected tissue continues increasing its size and ends forming a protuberance or 'bump', similar to a ball. Usually, initial symptoms occur during the formation of young developing fruits, in which latex can be observed bleeding over fruit epidermis and peduncle with an initial deformation that becomes evident very slowly (Nishina 1991). Seeds in affected fruits are often aborted or poorly developed and vascular tissue is mild and almost always with necrosis (Saran et al. 2013a, b, c, d). Under severe deficiency situations, height of trees may be affected causing a slight rosette effect and an associated stunting in apical part. In plants, the first symptoms are an abundant blooming, a latex secretion during fruit development and a deformation of these fruits later on (Figure 11.1). If soil analysis does not indicate boron availability for the plant, for this reason there has to be considered the deficiency symptoms. Many times, deficiency is confused with the disease of 'Bunchy Top', and for this reason it is suggested to make the test of fruit chopping and observe latex secretion. Boron deficiency appears more frequently in shallow, sandy and rocky soils in dry conditions. Trees with deformed fruits typically have boron levels in petioles (dry weight basis) of about 20 ppm and below (Wang and Ko 1975). Normal boron levels in petioles are about 25 ppm and higher (Chan and Raveendranathan 1984).

TABLE 11.1
Screening of Germplasm against Bumpiness and Market Acceptability of Affected Fruits

Germplasm Line	Incidence (%)	Boron Requirement[a]	Market Acceptability (Rs/kg)	
			Normal	Bumpy
Pusa Dwarf	22.22	Low	20	15.88
Pusa Red selection	04.17	Low	20	15.88
Pusa Nanha	05.56	Low	20	15.88
Pusa Delecious	12.00	Low	20	15.88
Pusa Majesty	14.29	Low	20	15.88
Co-6	35.29	Medium	20	14.00
Pune Selection-1	41.67	Medium	20	14.00
Pune Selection-2	45.45	Medium	20	12.00
Pune Selection-3	64.52	High	20	08.25

Source: Adapted from Saran, P. L., R. Choudhary, I. S. Solanki and P. R. Kumar, 2014a, *African Journal of Biotechnology,* 13(4): 574–80; Saran, P. L., R. Choudhary and I. S. Solanki, 2014b, *An International Event on Horticulture for Inclusive Growth,* Tamil Nadu, India: Tamil Nadu Agricultural University Coimbatore, p. 154; Saran, P. L., R. Choudhary and I. S. Solanki, 2014c, *Agriculture Today,* 17(2): 40–41.

[a] Low ≤ 25%; Medium = 26%–50%; High = 51%–75%.

FIGURE 11.1 Bumpy fruits and deformed seeds of papaya.

11.1.2 Management

Among micronutrients, boron is the most important in papaya crop, affecting directly the quality and production of fruits. Deficiency is caused by an excessive acidity, water stress and lower contents of organic matter and boron in soil. The ameliorative effect of boron application (as disodium octaborate tetrahydrate) through foliar sprays (0.1%) is more effective than soil application. The basal application of borax at 5.0 g/plant at onset of reproductive stage is very effective to control the bumpy fruit disorder of papaya, however, application of 13% borax at 12.5 kg ha^{-1} or boron at 1.63 kg/ha for sandy loam soils of North Bihar was also found effective (Saran et al. 2013a, b, c, d). It is important to go for only one spray of boron at initial fruit development stage and if both are combined, it may prove toxic to the plant. It is also suggested to eliminate through pruning all the fruits before physiological ripening that show bumpiness aspect and boron deficiency, because they do not have commercial value due to bad appearance and low level of sweetness in its pulp.

11.2 CARPELLOIDY FRUITS

Papaya plants in orchards, sometimes, fail to develop appropriate fruit shape (Figure 11.2). The plant may begin to develop fruits, but they are deformed/warty. Gynodioecious lines were the sensitivity of many hermaphrodite genotypes to fruit deformity caused by stamen carpelloidy. This is a genetic proclivity affecting floral development in hermaphrodites such that stamens become carpel-like and attach to the ovary in irregular or occasionally symmetrical lobes particularly during the cool, wet season. Carpel abortion is usually most pronounced under warm, dry season (Manshardt 2012).

The hermaphrodite plants are sexually ambivalent, producing staminate, perfect and pistillate flowers. The pistillate plant is stable. Staminate and hermaphrodite plants may be phenotypically stable or ambivalent, going through seasonal sex-reversal, during which they produce varying proportions of staminate, perfect and pistillate flowers (Ram 1996). The female plants normally bear pistillate flowers, but rarely, they can produce bisexual flowers. Alterations and reversions of bisexuality have frequently occurred during angiosperm evolution and resulted in functional unisexuality (male or female sterility) or morphological unisexuality (Mitchell and Diggle 2005). Through these alterations, carpelloidy/pistilloidy, restricted to the

FIGURE 11.2 **(See colour insert.)** Carpelloid fruits of hermaphrodite papaya line, PS-3.

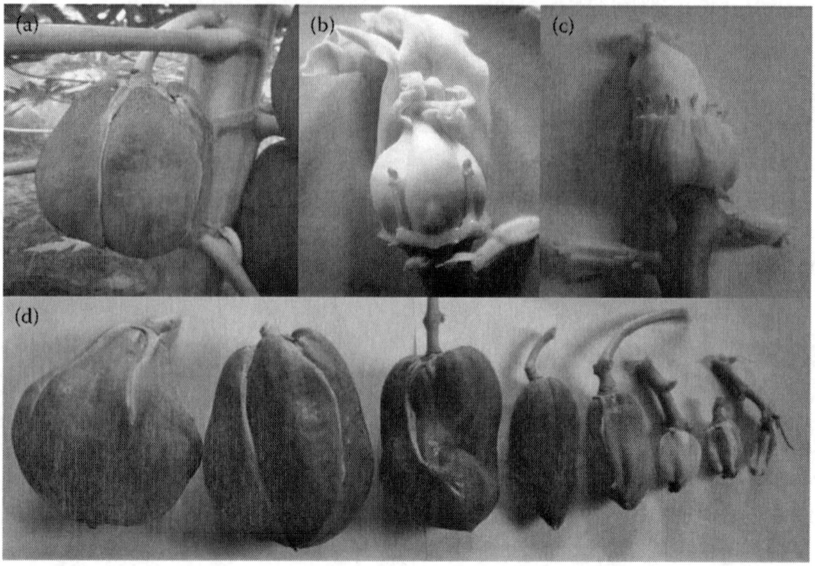

FIGURE 11.3 Carpelloid fruits bearing papaya plant (cv., Red Lady) (a), pistilloidy (stamen feminisation) (b), hermaphroditic flower (c) and different stages of stamen to carpel conversion in complete carpelloid fruit (d).

process of feminisation of stamen (Hama et al. 2004; Drea et al. 2007). Such type consists of bisexual flower types with stamen-to-carpel conversions. Hermaphrodite flowers of *Carica* tend to be highly variable in the extent of development of stamens and carpels, ranging from flowers with ten stamens and five carpels (elongata type) (Figure 11.3c) through abnormal flowers with 1–5 stamens fused to different degrees with the ovary and with some developing stigmatic tissue on the anthers (Figure 11.3b). The conversion process occurs rapidly upon stamen–carpel contact and all stamen tissues undergo feminisation (Figure 11.3a and d) due to mutation, gene alteration or environmental stress. The genetic nature of carpelloidy remains to be understood, possibly within the framework of the genetic system for sex determination which rests on five pairs of genes occurring in three sex-determining complexes in the sixth chromosome. The stamen filaments and connective tissues were the most responsive to feminisation. The *Carica* results highlight mechanisms that allow direct resource reallocation (Wright and Meagher 2003) between male and female organs through (partial) sex conversion once bisexual flowers have evolved.

Imperfect flowers' occurrence in hermaphrodite plants of papaya tree is related to genetic causes, which are affected by environmental factors. Longer dry periods, high temperature, high and low moisture (Manshardt 2012) and imbalanced fertilisation, may lead to this disorder. High humidity conditions as well as high concentration of water and nitrogen in the soil tend to change the sex of the hermaphrodite flowers producing deformed fruits.

The high nitrogen in papaya may be a factor behind low level of some micronutrients (especially boron). Any type of stress may also cause the problem as well as gene alteration through mutation. It was the highest in summer and become worse with the water deficit. The hermaphrodite varieties when exposed to very low and high temperature, high humidity and nitrogen level, the male parts (stamens) transform to carpel-like structure and form severely deformed fruit. The adoption of irrigation amount of 120% of evapotranspiration (ET) minimised the losses caused by production of imperfect flowers.

11.3 LUMPY FRUIT

Hard portions of tissue are frequently encountered in the flesh of ripe papaya fruits. This disease of unknown origin, found in orchards on the island Oahu (Hawaii) and subtropical parts of India, is referred to as 'lumpy' fruit. There are three types of symptoms, namely, large plate like areas in the fleshy portion of the fruit, small grain like lumps and rounded hemi-spherical lumps attached to the rind (Eloisa et al. 1994). Lumps can be artificially induced in papaya fruit by injections of a variety of chemicals including water, indole acetic acid (IAA) and maleic hydrazide (MH), or by physical injury. The presence of lumps in fruit can only be determined after the fruit is fully ripened.

11.4 VIVIPARY AND WHITE SEED

Vivipary and white seed (disturbed sarcotesta) are major seed production constraints of papaya in subtropical fringes of India. The precocious germination of seeds was observed in fruits of papaya cv., Madhubindu while they were still attached to the parent plant. The natural occurrence of precocious germination has been reported in papaya (Balakrishnan et al. 1986). These disorders are emerging physiological anomalies due to temperature fluctuation during seed and fruit maturity stage. Morphologically, the normal fruits and seeds are entirely different from disordered fruits (Figure 11.4).

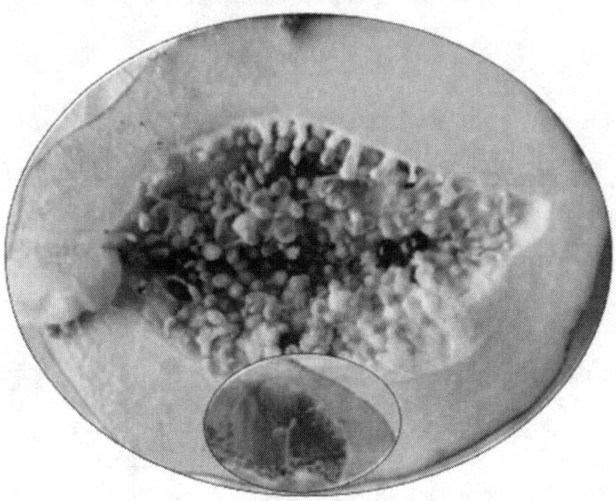

FIGURE 11.4 **(See colour insert.)** Vivipary in cv., Pusa Dwarf.

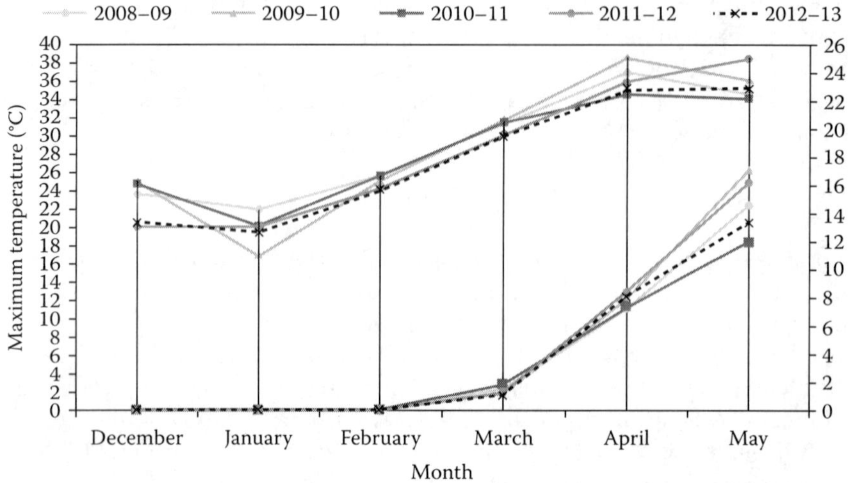

FIGURE 11.5 Relationship between monthly maximum temperature and vivipary in cv., Pusa Dwarf.

Average deformed seeds/fruit (+++) was significantly higher rated in February harvest as compared to fruits harvested in other months, while average vivipary seeds/fruit (+++) was observed significantly higher during May harvested fruits. The maximum incidence of vivipary and white seed disorders were observed during May (13.4% and 75.7%, respectively), while least incidence was observed in February and March (10% and 5%, respectively) (Figure 11.5). The highest economic loss (Rs. 53,600/ha) was observed in April, while the lowest during February (Rs. 16,800/ha) in autumn sown crop of Pusa Dwarf for seed production (Saran et al. 2013a, b, c, d) (Figure 11.6). *In vitro* culture study was conducted for maintenance or conservation of such native diversity (Saha 2007).

11.4.1 Management

Papaya is a tropical crop and requires 23°C average daily temperature for 6 months from flowering to fruit maturity. However, fluctuation of temperature, especially higher temperature in subtropical areas of India is very high, so it is suggested to avoid the susceptible cultivars and new plantation should be raised of resistant/tolerant cultivars under tropical area. Further studies are required to determine the exact causes and management of this disorder.

11.5 FROST DAMAGE

Low temperature is one of the most important environmental factors, which affect the plant growth and productivity. The winter temperatures will never go below 7°C during December to February but temperatures can vary greatly depending on regions. The leaves on affected plants start drying and petiole remain green; some time (Figure 11.7).

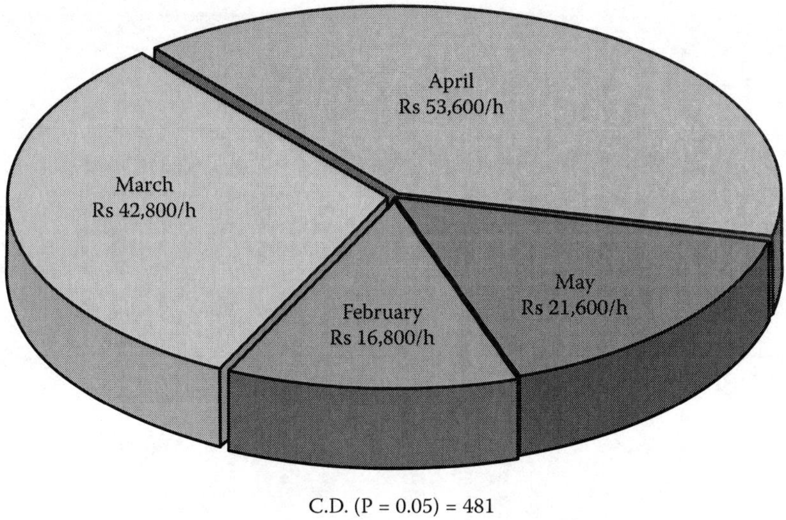

FIGURE 11.6 Economic losses due to viviparous seeds in cv., Pusa Dwarf.

FIGURE 11.7 Papaya plants damaged by frost.

11.5.1 Management

If temperatures at night are going very low, a temporary structure should be erected to save the plants in kitchen garden. This can be provided by building a frame around the plants and covering it with thatching material when frost threatens. Potted specimens can be moved to a frost-secure area. It is important to cover the plant all the way to the ground to trap heat radiated from the ground at night. The new plantation of papaya can be saved from frost by covering the plant with polythenes (Figure 11.8). During the month of January, regular light irrigation combined with smoking through crop wastage or stubbles in the evenings help in avoiding the frost injury.

FIGURE 11.8 Protection of papaya seedlings from frost through polythene bags during winter.

11.6 CHILLING AND HEAT INJURY

Chilling injury symptoms include pitting, blotchy colouration, uneven ripening, skin scald, hard core or hard areas in the flesh around the vascular bundles, water soaking of tissues and increased susceptibility to decay. Increased *Alternaria* rot was observed in mature-green papaya kept for 4 days at 2°C, 6 days at 5°C, 10 days at 7.5°C or 14 days at 10°C. Susceptibility to chilling injury varies among cultivars and is greater in mature-green than ripe papaya (10 vs. 17 days at 2°C; 20 vs. 26 days at 7.5°C).

Heat injury exposure of papaya to temperatures above 30°C (86°F) for longer than 10 days or to temperature-time combinations beyond those needed for decay and/or insect control result in heat injury (uneven ripening, blotchy ripening, poor colour, abnormal softening, surface pitting and accelerated decay). Quick cooling to 13°C (55°F) after heat treatments minimises heat injury (Kader 2000).

11.7 SKIN ABRASIONS

Skin abrasions result in blotchy colouration known as green islands. In this, skin remains green and sunken when fruit is fully ripe and accelerate water loss. Abrasion and puncture injuries are more important than impact injury.

12 Unfruitfulness

Hardly a day goes by without farmers calling to ask why their papaya trees are not bearing/fruiting, or are dropping and diseased. Sometimes, seedless papaya (parthenocarpic) fruits are found in the market. Seedless fruits develop from the pure female or hermaphrodite flowers which do not get fertilised by the pollen from male or hermaphrodite trees. This type of fruit is born, generally, in the beginning of fruiting season when there is lack of pollen. Seedless fruits are, generally, smaller in size and very sweet. When papaya plants bear no fruiting or sparse fruiting takes place in the orchard, it is called as unfruitfulness (Ram 1983). If, the following points are kept in mind, this malady can be remedied.

12.1 GENETIC INFLUENCES

Gynoecious and andromonoecious varieties are preferred because they consist of female and hermaphrodite plants that produce marketable fruit for fresh consumption. These populations are derived from crosses between hermaphrodite plants in a ratio 2:1 in segregation of female and hermaphrodite plants. This is somewhat advantageous for producers, since only the hermaphrodite plants remain in the plantations after the definitive sexing procedure, once consumers prefer fruits of hermaphrodite plants (Khan et al. 2002). Recent molecular studies have shown that sex determination in this species is controlled by a recently evolved homomorphic pair of sex chromosomes (type X/Y), differentiated by a small male-specific region on the Y chromosome (MSY). In this sense, the female plants are homogametic with the XX chromosomes, whereas male and hermaphrodite plants are heterogametic, with XY and XYh chromosome combinations, respectively (Liu et al. 2004). It is likely that two genes are involved in papaya sex determination, the first gene is a suppressor of the stamen in female flowers (feminising gene) and the other gene is a suppressor of carpel in male flowers (masculinising gene) (Ming et al. 2007). More recent research has detected seven genes in the sex-controlling region, which were however not able to differentiate the three sex types, because there was neither differential expression nor dosage effect, suggesting that these genes are not involved in sex determination (Yu et al. 2008a, b). Despite certain progress in research on the molecular mechanisms of papaya sex determination, there is little understanding of the expression of the sexual forms and variations, which are directly related to the production efficiency of marketable fruit. This issue is considered complex and intriguing in view of the lethal factor associated with the dominant alleles responsible for hermaphroditism and masculinity (Ming et al. 2007), aside from the influence of genetic and environmental factors on both (Damasceno Junior et al. 2008). The combination of these factors is possibly responsible for the high degree of instability of hermaphrodite plants in terms of sex expression; a variation of flowers to carpelloid and pentandric forms or sex reversal is possible. These variations of the hermaphrodite flower, classified as floral abnormalities, reduce crop yields and increase seasonality in fruit yield, leading to supply oscillations and consequently to price variations of papaya on the market. Carpelloid fruits occur due to the transformation of stamens into carpel-like structures, producing fruit with varying degrees of malformation, while pentandrya is the transformation of hermaphrodite into a typically female flower, with a reduced number of stamens and an ovary with five deep grooves. These variations occur mainly at mild or low temperatures and high moisture and soil nitrogen levels (Awada 1958; Arkle Junior and Nakasone 1984) and are undesirable in plantations, since their fruit has no market value, decreasing yields. Sex reversal, however, is a result of abortion or ovarian atrophy, producing no fruit. This phenomenon occurs related to high temperatures, water stress and low soil nitrogen (Awada 1958; Arkle Junior and Nakasone 1984).

12.2 ENVIRONMENTAL INFLUENCES

Papaya flourishes in the frost-free and humid areas of the tropics and subtropics with an average temperature of 25°C and rainfall between 1500 and 2000 mm, which spread throughout the year. Due to its soft wood and flat root system, the papaya is very susceptible to wind break, especially in a monoculture. Soil must be well drained because papayas are susceptible to stagnant water. Dry periods longer than two months may lead to the necessity of irrigation. It reacts very sensitively to cultivation mistakes (Naturlande 2000). Even under optimum sites, plantations have encountered several problems that have brought low production due to poor varieties, weediness, pollination, abiotic and biotic stress, etc., as given below:

12.2.1 Choice of Variety

Heavy or shy bearing is a genetical character in papaya. But some acclimatised varieties give good yield in a specific zone. The cultivar Ranchi is acclimatised to Jharkhand conditions and has a wide adaptability under North Indian conditions. Coorg Honey performs well in South India, especially in Karnataka but a poor yielder in North India. Similarly, Barwani in Central Zone, Washington in Western Zone and Co varieties in a good part of Southern zone perform well. Exotic varieties like, Solo, Sunrise Solo, Waimanalo, Taiwan and Thailand perform well in Southern and Western India but give poor performance under North Indian conditions. Homesteads from Nigeria show good potential in yield all over the country. Pusa varieties have also shown good performance under most parts of the country. Therefore, only tested varieties suitable for a specific zone should be grown.

In the dioecious condition, the male and female plants are separate, so out-crossing is requisite to fertilisation. In a dioecious variety (Washington), anther dehiscence was completed 36–18 h before anthesis and stigmas became receptive a day before anthesis (Khuspe and Ugale 1977). In gynodioecious varieties, self-fertilisation is possible in hermaphrodite flowers. Anthers dehisce before anthesis, facilitating cleistogamy (Chan et al. 1999). Maximum stigma receptivity has been found to occur on the day of anthesis (Dhaliwal and Gill 1991). A hermaphrodite flower's pollen may be released before its stigmas' are receptive (protandrous dichogamy), with the stigmas becoming receptive only at anthesis (Pares et al. 2002). There may be genotype differences in variety lines, and seasonal changes in flower receptivity affecting pollination (Pares-Martinez et al. 2004). In gynodioecious plants, seed set was 10 times greater when Coorg Honey Dew plants in India were hand-pollinated after being open-pollinated (Purohit 1980), but hand pollination did not increase papaya fruit set. These results depend upon the pollinators available as well as the papaya variety (Rodriguez-Pastor et al. 1990).

12.2.2 Planting Season

Papaya is a tropical fruit and it requires warm and humid season. Winter season retards the growth and fruiting of this crop. Most parts of our country are subtropical. The papaya starts flowering and fruiting after about 6 months under normal condition. Therefore, planting should be manipulated in such a way that fruiting may continue for a longer period. The optimum period has been found to be August–September for sowing seeds in the nursery and October–December for planting. This provides an extended long range of warm and humid condition for fruiting (Ram and Ray 1992). Planting done in other season either takes a longer time for fruiting or leads to sparse fruiting in the first season.

Under commercial conditions, papaya has exacting climate requirement for vigourous growth and fruit production. It must have warmth throughout the year and will be damaged by light frost (O'Hare 1993), even brief exposure to 0°C is damaging and prolonged cold without overhead sprinkling will kill the plants (California Rare fruit Growers Inc 1997). Temperatures below 12–14°C strongly retard fruit maturation and adversely affect fruit production (Nakasone and Paull 1998) but hotter temperatures, if accompanied by dry conditions, can also adversely affect fruit set (Elder et al. 2000). Similarly, soil temperatures below 15°C limit growth (O'Hare 1993). An ambient temperature range between 21°C and 33°C is ideal (OECD 2005).

12.2.3 Organic Manuring

Basal dose of organic manure should be applied in pits and not flat beds. In a field trial, cv., Pusa Delicious was transplanted in (a) pits dug 90 days and filled with manures 60 days ahead of the transplanting date, (b) pits dug and filled with manures 2–3 days ahead of the transplanting and (c) in flat lands (beds) prepared 2–3 days ahead by deep ploughing after broadcasting the manures in a metre-wide band running along the planting row. Plants getting treatment (a) flowered quite earlier than those under treatment (c) and produced the highest fruit yield, 36.40 kg/plant. The result with (b) was statistically comparable to that of 34.50 kg/plant (a). Transplanting in flat land (c) brought about significant reduction in the plant productivity (9.1 kg). These observations give strong support for transplanting papaya in pits dug and manured either recently or in advance in heavy soil (Ram and Ray 1992).

12.2.4 Nutrition

Papaya is a heavy-feeder fruit crop. If sufficient amount of balanced fertiliser is applied to the plants, they fail to produce fruit or give meager fruiting. Mycorrhizal interactions of male and female papaya plants may differ: females seem more responsive to changes in soil fertility and readily adjust mycorrhizal colonisation accordingly (Vega-Frutis and Guevara 2009). Many studies are reported on the positive effects of arbuscular mycorrhizal fungi on papaya (Sukhada 1992). Judicious use of organic manures, inorganic and microbial fertilisers is rapidly gaining favour in fruit crops (Anwar et al. 2005). It is also necessary to strike a balanced (Carbon: Halogen) ratio for fruiting properly. Besides, some micronutrients also play a vital role in fruiting. Boron is also essential for pollination and seed production (Gupta et al. 1985). Other soils are naturally deficient in certain elements, making it necessary to amend them with fertilisers (Nelson 2012a, b). Lack of moisture over a prolonged period will slow down the growth and encourage the production of a number of male or sterile flowers. The result is that fewer fruits are set on the tree resulting unfruitfulness in papaya. Low moisture levels or low nitrogen can induce female sterility (Awada and Ikeda 1957). Abiotic stresses like, low temperatures, high soil moisture and high nitrogen seem to produce carpellodic flowers (da Silva et al. 2007; Jimenez et al. 2014).

12.2.5 Pollination

Pollination is required for good setting of fruits and seed production. The farmers generally remove all the male plants, considering them to be unproductive, and because fruiting cannot take place in absence of pollination. For proper fruiting, pollination is essential and for adequate pollination, sufficient pollinisers are required. In dioecious varieties, 10%–12% male plants in the orchard population are required for proper pollination (Ram 2005).

Cross-pollination may be common or infrequent, depending upon the papaya variety, flowering behaviour and the environment. In some instances, male plants may more effectively pollinate hermaphrodites in adjacent orchards than the hermaphrodites can self-pollinate. Pollen can be produced year-round. The grains are relatively large (32–39 μm), and in the subtropics can be larger in local warmer areas (Sippel and Holtzhausen 1992). Viability of pollen may vary seasonally, being highest in the rainy season and spring, while the lowest in winter in subtropical (OGTR 2003). The flowers open in the early night-time (Pares et al. 2002), or the morning (Azad and Rabbani 2004), and since they are strongly dimorphic or polymorphic, provide different cues to potential insect pollinators. Staminate flowers may be more fragrant and open for 24 h, and they produce calcium oxalate crystals in the anthers and nectar basally, thus being an attractant for insects. The pistillate flower has no nectar, but a sweet non-sugar exudate seems available on its flared large antler-like stigmas (Pares et al. 2002). The female flowers may remain open for seven days (Mabberley 1998).

Although the floral morphology in papaya plants suggests insect pollination, various authors have indicated that wind pollination may also be important (da Silva et al.

2007). Honeybees, thrips, hawk moths and apparently also mosquitoes and midges have been reported as pollinators of papaya (OGTR 2003). In some countries, the role of insects in papaya pollination is factored as prominent, whereas in others wind-borne pollen appears to be of more concern. Accordingly, different recommendations for appropriate isolation distances from other papaya may reflect the specific conditions at different locations of production. Recognising both insects and wind as agents for pollen movement, Singh (1990) recommended 2–3 km isolation for production of foundation seed, but cited no experimental observations supporting this distance.

12.2.6 Weed Control

Weeds grow luxuriantly in the papaya orchard and exhaust most of the nutrients applied to it. They also compete for light, air and water, which results in poor fruiting in papaya. Even after repeated weeding, they come up regularly. Best way of manuring is to apply basal dose in pits of 60 cm × 60 cm × 60 cm and then go for planting because at this depth, the grasses are generally unable to share the nutrients. The fertilisers in the form of top-dressing should be applied around the covered soil in the root zone. Weeding is done regularly, especially around the plant, so that the nutrients applied are not taken up by the weeds. Herbicide usage is the most sustainable practice and, when compared to other strategies such as hoe weeding and intercropping, had beneficial effects on commercially important parameters in terms of reduction in days to 50% flowering and an increase in fruit yield in variety 'Sunrise Solo' (Akinyemi et al. 2004).

12.2.7 Water Requirement

Papaya is a crop that cannot withstand water logging but needs plenty of water, especially when it starts fruiting. However, this requirement is fulfilled in the rainy season. If there is drought or scanty rain, irrigation must be provided copiously. From October to April, when the upper fruits are still developing, the plant should be irrigated frequently. Irrigation can also be given after top-dressing in the rainy season, if the rain fails. In areas of high rainfall, B leaches through the soil profile and out of the root zone. In acidic soils, certain elements may become bound to clay particles and be unavailable to plants. Some soils are compacted, lacking in oxygen, and need to be aerated. Papaya plants are tolerant to water logging under well-drained soil conditions (Marler et al. 1994; Benson and Poffley 1998). Waterlogged plants die within two to four days due to stomata closure and abscission of expended leaves (Marler et al. 1994). Plants that do not die, do not recover well.

12.2.8 Intercropping

First year of planting, sufficient space is available in between two rows of papaya plantation, in which intercrops can be grown successfully. The crops selected should have no adverse effect on papaya. The crops that compete with papaya or suppress its growth should not be included in the intercropping. Virus susceptible crops (tomato, brinjal and okra) normally should not be grown. Leguminous crops of low height like, lentil, gram, greengram and blackgram should be preferred, because they add nitrogen into the soil and suppress the weeds.

12.2.9 Crop Rotation

A suitable crop rotation is necessary for a viable papaya production. Taking continuously the same crop year after year results in the complete exhaustion of some essential macronutrients and micronutrients. Besides, it increases the intensity of pests and diseases, which leads to poor production of the fruits. Intercropping of leguminous crop after non-leguminous crops and shallow-rooted crops after deep-rooted crops removes this problem. A few suitable crop rotations in India are

1. Papaya (October–May), maize (June–September), wheat (November–April) and maize (June–September).
2. Maize (June–September), papaya (October–May), green-manure crop (June–September) and wheat (November–April).
3. Sugarcane (February–February), greengram (March–June), maize (June–September), papaya (October–May), green-manure crop (June–September) and Potato (November–February).

12.2.10 Diseases

Plants heavily infected with viral diseases either fails to produce fruits or give markedly reduced yield. As a result, papaya was wiped out between 1950 and 1980 in NEPZ of India. The juvenile papaya seedlings which have least faced the period of rainy season are generally free from this disease. Therefore, seeds should be sown in the nursery in the beginning of September and crop should be planted in October, so that it escapes rain in the first year. The plants become hardy in the ensuing winter and summer seasons. Thus, there is little infection of viruses in the next rainy season and fruiting in papaya is at its maximum (Ram and Ray 1995).

Humid and heavy rainfall area is very favourable for the spread of root-rot disease. The tap root or deeper roots are found rotten. This is owing to over-saturation of water in the subsoil. The uptake of nutrients by the superficial fibrous roots is partial; as a result there is no fruiting or poor fruiting in papaya. Therefore, the area having very high water-table should be avoided. However, the up and sloppy land in such an area may be utilised for papaya planting. If at such places there is danger of root-rot, 100 g copper sulphate and 1.0 kg lime should be added along with organic manures in the pits as a basal dose which is very effective. The long-term applications of inorganic fertilisers not only decrease the crop productivity (Chand et al. 2006), but also make the plants susceptible to insects and diseases, besides deterioration of fruit minerals and quality.

12.2.11 Plant Growth Regulators

Papaya trees are fast-growing and prolific and can often result in widely separated internodes; the first fruit is expected in 10–12 months from germination and in general the fruit takes about 5 months to develop. Soil application of paclobutrazol, a growth retardant, at 1000 mg/L resulted in reduced overall height and reduced height at which first flowers bud; it did not affect the start of production or yield (Rodriguez and Galan 1995). Spraying of gibberellic acid (200 ppm) on female flowers of cv. Honey Dew is helpful in producing seedless fruits. GA_3-treated fruits contained only 10–87 seeds compared with 500–800 seeds in untreated fruits. The Vitamin C content was also doubled (Ram 1983).

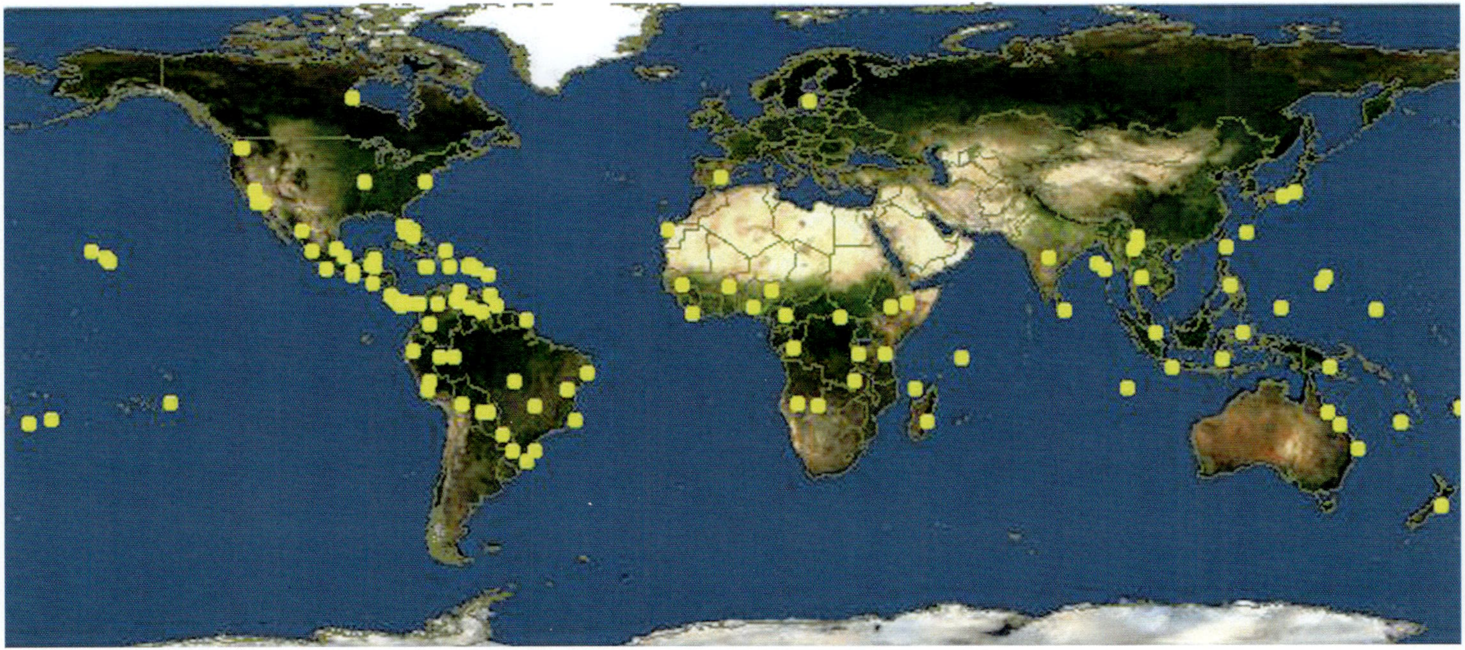

FIGURE 1.2 Major papaya growing countries. (From Encyclopaedia of Life (EOL), 2015, http://eol.org/pages/585682/maps.)

FIGURE 1.5 Mature papaya tree cv., Pusa Dwarf.

FIGURE 2.2 Major floral visitors/pollinators of papaya.

FIGURE 2.3 *V. cauliflora* tree in fruiting.

FIGURE 2.4 Papaya (*C. papaya*) flowers: (a) male flowers, (b) male flower in longitudinal section, (c) female flower, (d) hermaphrodite flower, (e) hermaphrodite flower of pentandria type, (f) hermaphrodite flower with carpelloid stamens.

FIGURE 2.6 Important papaya varieties from IARI: (a) Pusa Majesty, (b) Pusa Delicious, (c) Pusa Giant, (d) Pusa Dwarf and (e) Pusa Nanha.

FIGURE 2.7 Important papaya varieties from Tamil Nadu Agricultural University, Coimbatore: (a) CO-1, (b) CO-2, (c) CO-3, (d) CO-4, (e) CO-5, (f) CO-6 and (g) CO-7.

FIGURE 3.1 Raised beds for papaya plantation.

FIGURE 4.1 Heavy-bearing papaya needs staking to prevent lodging.

FIGURE 4.2 Papaya fruit size: (a) after thinning and (b) without thinning.

FIGURE 4.4 Position of sprouting buds in one-year-old pollard papaya plant.

FIGURE 4.5 Papaya on raised bed with intercrops: (a) turnip, (b) lentil and (c) suran.

FIGURE 4.6 Collar rot disease–free papaya on raised bed.

FIGURE 5.1 (a) Multiple shoot induction on MS medium with BAP (5 μM) + NAA (0.05 μM), (b) elongation on half MS with 1.5 μM of GA_3; (b1) stunted shoot before GA_3 treatment, (b2) elongated shoot after GA_3 treatment, (c) rooting of *in vitro* regenerated shoots in half MS with 2.5 μM of IBA, (d) plant with well-developed root system ready for hardening, (e) plant acclimatised to greenhouse conditions. (Adapted from Anandan, R, S. Thirugnanakumar, D. Sudhakar and P. Balasubramanian, 2011, *Journal of Agriculture Technology*, 7(5): 1339–48.)

FIGURE 5.2 Nursery raising: seed sowing (a), seedlings on raised beds (b) and seedlings in poly tubes (c).

FIGURE 6.1 Fertiliser application in papaya.

FIGURE 6.2 Bumpy fruits of papaya.

FIGURE 6.3 Symptoms of boron toxicity on the leaves.

FIGURE 8.1 Traditional method of seed extraction.

FIGURE 8.2 Deformed and discoloured seeds of papaya.

FIGURE 9.1 Mealy bug infestation on papaya plants.

FIGURE 9.2 (a) Female fruit fly and (b) damaged fruit. (Adapted from Pena, J.E. and F.A. Johnson, 2006, *Insect Management in Papaya*, The Institute of Food and Agricultural Sciences (IFAS), University of Florida, ENY-414:1–4.)

FIGURE 9.3 Yellow panel trap for female fruit fly.

FIGURE 9.4 Papaya fruit infested with female scale insects. (Adapted from Pena, J. E. and F. A. Johnson, 2006, *Insect Management in Papaya*, The Institute of Food and Agricultural Sciences (IFAS), University of Florida, ENY-414:1–4.)

FIGURE 9.6 Snails (a) and crows (b, c) eating seeds and fruits during winter in a papaya orchard.

FIGURE 9.7 Symptoms of phytotoxicity.

FIGURE 10.2 Mosaic symptoms on papaya plant.

FIGURE 10.4 (a) Root tip, (b) root rot initial stage and (c) later stage.

FIGURE 10.8 Fruits damaged by anthracnose disease: (a) unripe and (b) ripe stage.

FIGURE 11.2 Carpelloid fruits of hermaphrodite papaya line, PS-3.

FIGURE 11.4 Vivipary in cv., Pusa Dwarf.

FIGURE 14.1 Papaya fruits with the sign of ripening: (a) Pusa Majesty and (b) Pusa Dwarf.

13 Papain

Papain is an endolytic plant cysteine protease enzyme which is isolated from papaya (*C. papaya* L.) latex. Laticifers of papaya are complex tissue systems of the articulate-anatomizing type. These conducts are multi-cellular columns with perforated transverse lateral walls, protoplast fusion, and intrusion of phloem cells, forming branched networks (Hagel et al. 2008). Damage to any aerial part of the papaya plant, where laticifers are widely distributed, elicits latex release, which is very typical for this species (Azarkan et al. 2003).

The papain present in the latex is an enzyme of industrial use and of high research interest (Saran and Choudhary 2013). Papain exporting countries are Zaire, Tanzania, Kenya, Israel, Philippines, Sri Lanka, Cameroun and India and the major source of spray dried papain is Zaire. Principal importers are the United States, Japan, the United Kingdom, Belgium and France. Other countries import papain in small quantities. Papain is in great demand in international market, particularly the United Kingdom, the United States and Europe (Ram 2005). With the increase in the cultivation of papaya in various parts of India, it is highly profitable to manufacture papain where fruit production is in abundance (Ram 2005). In India, papain can be supplied and sold to Enzo-Chem Laboratory Yeola, Nasik and M/s. Pharmaceutical and Cosmetic Promotion Council, Mumbai.

Almost all the best quality papain goes to the England and United States for use as crude papain in the brewing industry for chill-proofing beer. However, the increasing trend for additive free beers initiated by other European countries is taking effect in England and so this market for papain is declining. Among the major applications, it is used in the food industry (Saran and Choudhary 2013), beer clarification, meat tenderising, preparation of protein hydro-lysates, etc. The lanced fruits may be allowed to ripen and can be eaten locally, or they can be employed for making dried papaya 'leather' or powdered papaya, or may be utilised as a source of pectin. Because of its papain content, a piece of green papaya can be rubbed on a portion of tough meat to tenderise it. Sometimes, a chunk of green papaya is cooked with meat for the same purpose. One of the best known uses of papain in commercial market is as meat tenderiser, especially for home use. A modern development is the injection of papain into beef cattle half an hour before slaughtering to tenderise the meat than that it would normally be. However, the papain-treated meat should never be eaten raw but should be cooked sufficiently to inactivate the enzyme. The tongue, liver and kidneys of injected animals must be consumed quickly after cooking or utilised immediately in food or feed products, as they are highly perishable. Papain has many other practical applications; it is used to clarify beer, also to treat wool and silk before dying, to dehair hides before tanning, and it serves as an adjunct in rubber manufacturing (Amri and Mamboya 2012). It is applied on tuna liver before extraction of the oil which is thereby made richer in vitamins A and D. It is incorporated into toothpastes, cosmetics, detergents and pharmaceutical preparations to aid digestion.

13.1 COMPOSITION

This milky latex is a slightly acidic fluid composed of 80% water (Rodrigues et al. 2009). It contains sugars, starch grains, minerals (S, Mg, Ca, K, P, Fe, Zn), alkaloids, isoprenoids, lipidic substances and proteins, including enzymes such as lipases, cellulases and cysteine proteases (papain, chymopapain) which are important in defense against insect herbivores and in tissue and organ formation (pith differentiation) (Konno et al. 2004). In young fruits, laticifers develop near the vascular bundles and separate through transverse walls that later dissolve, making these laticifers a series of superposed fused cells.

13.2 PROPERTIES, STRUCTURE AND CHARACTERISTICS

The globular protein, the papain PDB accession number, 1CVZ, is a single chain protein with molecular weight of 23,406 Da and consists of 212 amino acids with four disulphide bridges and catalytically important residues in the following positions, Gln19, Cys25, His158 and His159 (Mitchel et al. 1970; Robert et al. 1974; Tsuge et al. 1999) (Figure 13.1).

Papain is a cysteine hydrolase that is stable and active under a wide range of conditions. It is very stable even at elevated temperatures (Cohen et al. 1986). Papain is unusually defiant to high concentrations of denaturing agents such as, 8M urea or organic solvent like, 70% ethanol. Optimum pH for activity of papain is in the range of 3.0–9.0 which varies with different substrate (Edwin and Jagannadham 2000; Ghosh 2005). Papain enzyme as cysteine proteases in papain superfamily is usually consisting of two well-defined domains which provide an excellent system for studies in understanding the folding–unfolding behaviour of proteins (Edwin et al. 2002). The protein is stabilised by three disulphide bridges in which the molecule is folded along these bridges creating a strong interaction among the side chains which contribute to the stability of the enzyme (Tsuge et al. 1999; Edwin and Jagannadham 2000). Its three-dimensional structure consists of two distinct structural domains

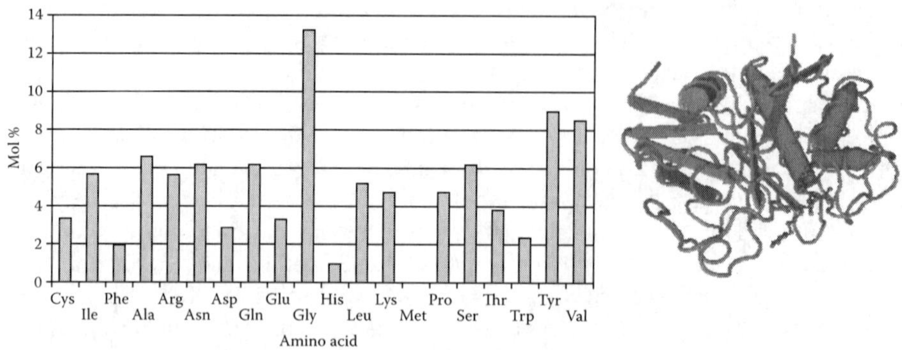

FIGURE 13.1 Graphical representation of the amino acid composition of papain (a); papain structure (MMDB protein structure, 1CVZ) (b). (Adapted from Amri, E. and F. Mamboya, 2012, *American Journal of Biochemistry and Biotechnology,* 8(2):99–104.)

with a cleft between them. This cleft contains the active site, which contains a catalytic diad that has been likened to the catalytic triad of chymotrypsin histidine-159. Aspartate-158 was thought to play a role analogous to the role of aspartate in the serine protease catalytic triad, but that has since then been disproved (Menard et al. 1990). Papain molecule has an all-α domain and an anti-parallel β-sheet domain (Kamphuis et al. 1984; Madej et al. 2012). The conformational behaviour of papain in aqueous solution has been investigated in the presence of SDS and reported to show high α-helical content and unfolded structure of papain in the presence of SDS is due to strong electrostatic repulsion (Huet et al. 2006). In the molten globule state (pH 2.0), papain shows evidence of substantial secondary structure as ß-sheet and is relatively less denatured compared to 6 M guanidium hydrochloride (GnHCl). The enzyme also exhibits a great tendency to aggregate at lower concentrations of GnHCl or a high concentration of salt (Edwin and Jagannadham 2000). Papain is often useful to examine the relative hydrophobicity or hydrophilicity values of the amino acids in a protein sequence (Amri and Mamboya 2012). The enzyme has been reported to be generally more stable in hydrophobic solvents, at lower water contents and can catalyse reactions under a variety of conditions in organic solvents with its substrate specificity little changed from that in aqueous media (Stevenson and Storer 1991). Hydrophobicity of papain enzyme being maintained at 31.45% of carbon all along the sequence contribute to stability of protein as has been previously reported that stable and ordered proteins maintain 31.45% of carbon all along the sequence (Jayaraj et al. 2009).

13.2.1 Mechanism

The mechanism in which the function of papain is made possible is through the cysteine-25 portion of the triad in the active site that attacks the carbonyl carbon in the backbone of the peptide chain freeing the amino terminal portion. As this occurs throughout the peptide chains of the protein, the protein breaks apart. The mechanism by which it breaks peptide bonds involves deprotonation of Cys-25 by His-159. Asparagine-175 helps to orient the imidazole ring of His-159 to allow this deprotonation to take place. Although far apart within the chain, these three amino acids are in close proximity due to the folding structure. It is through these three amino acids working together in the active site that provides this enzyme with its unique functions. Cys-25 then performs a nucleophilic attack on the carbonyl carbon of a peptide backbone (Menard et al. 1990; Tsuge et al. 1999). In the active site of papain, Cys-25 and His-159 are thought to be catalytically active as athiolate-imidazolium ion pair. Papain can be efficiently inhibited by peptidyl or non-peptidyl *N*-nitrosoanilines (Guo et al. 1996, 1998). The inactivation is due to the formation of a stable S–NO bond in the active site (S-nitroso-Cys25) of papain (Xian et al. 2000).

13.3 PRODUCTION TECHNIQUE

Papain extraction is very simple and easily oxidised by exposure to air, and is destroyed in aqueous solution by temperature above 70°C or by some light. It is also easily inhabited by contact with some metals such as iron, copper, zinc, etc., and works best at particular temperatures. Fresh latex is obtained from 75 to 90 days old developing green fruits fit for latex extraction. This starts in rainy season and continues up to March. Cool and wet periods produce more papain. It is obtained by making incisions on the surface of the green fruits early in the morning.

Maximum latex flow was obtained between 5:00 and 10:00 a.m. The incising knife is made using a stainless steel razor blade set into a piece of rubber attached to a long stick used for lancing the fruit. The blade should not extend beyond more than 2 mm as cuts deeper than 2 mm because juices and starch from the fruit pulp mix with the latex, which lowers the quality. Four vertical incisions are made in the fruits with a sharp knife to a depth of 1–2 mm on fruit from stalk end to the fruit tip (Figure 13.2). The incision should be repeated on 3–4 subsequent occasions at an interval of 4–5 days. A non-metallic container should be used for collecting the latex. The greener the fruit, more active is the latex flow. The higher the specific activity or rate of catalysing per unit weight of enzyme, higher is the value. The activity of papain can be increased by refinement in processing. The quality and the grade of papain is determined by colour and enzyme activity (tyrosine unit) which is classified in three groups, (i) crude papain white brown to brown, (ii) crude papain in flake or powder referred as semi-refined and (iii) spray dried crude papain of higher activity in powder form referred to as refined papain. The refined papain having the activity of 800–1000 units or higher is most preferred in world trade. Extraction of latex must tap more than 1000 average-size fruits to gain 450 g of papain. Fruits should be tapped at intervals of about 4–7 days and for the first tapping it is usually sufficient to make only one cut. On subsequent tapping the two or three cuts are spaced between earlier ones.

FIGURE 13.2 Papain extraction technique.

The latex after incision takes only 1 to 2 min for coagulation and can be collected into a glass container. Tappers also hold containers, like an 'inverted umbrella' clamped around the stem, which may be coconut shell/clay cup/glass/porcelain/enamel pan beneath the fruit to collect the latex. Sometimes, the fruits are also harvested at unripe stage and latex is extracted. After extraction, the latex is immediately used for the purification of papain in its native state or stored at −8°C. The latex coagulates quickly and for best results, it is spread on fabric and oven-dried at a low temperature, then ground to powder, and packed in tins. Sun-drying tends to discolor the product.

13.3.1 Manufacturing Process

Papaya latex is mixed with potassium meta-bi-sulphate (KMS), sieved to remove extra materials and then blended with activated Zeo Carb cation exchange resin. The mixture is then centrifuged to separate out the resin from till liquefied latex. The latex is again mixed with activated anion exchange resin, blended and centrifuged. The liquefied latex is spread in trays and dried in vacuum shelf drier and mixed with potassium meta-bi-sulphate, ground and diluted with lactose to get IP grade papain.

13.4 PAPAIN CONCENTRATE

Papaya latex is mixed with potassium meta-bi-sulphate and distilled/de-mineralised water. It is then stirred thoroughly and stored in a cold room for settling. The supernatant liquid is siphoned off, filtered and the clear filtrate is mixed with denatured spirit and allowed to settle in cold room. The supernatant liquid is decanted from the slurry and the slurry is filtered. The filtered cakes are washed with rectified spirit and dried first with compressed air and then in vacuum self-dryer.

13.5 DRYING METHODS

After collecting the latex, it should be dried in electric oven at 40°C. Delay in the drying affects the quality of papain. Tapping and collection procedure, method of drying, storage and packing determines the quality of papain. The method of drying is the main factor that determines the final quality of papain. There have been various grades used since papain became an international commodity. Up to the mid-1950s when papain from Sri Lanka dominated the market three grades were known: (1) fine white powder, (2) white oven-dried crumb and (3) dark sun-dried crumb. Up to the 1970s there were two grades, namely, high grade oven-dried papain in powder or crumb form usually creamy white in colour and second or low grade sun dried brown papain in crumb form. Since 1970 as a result of new processing techniques, papain has been re-classified into three groups, namely, (1) crude papain – ranging from the first grade white down to second grade brown, (2) crude papain in flake or powder form, sometimes, referred to as semi-refined and (3) spray dried crude papain in the powder form, referred to as refined papain. There are different types of drying methods and driers for papain drying as given in Table 13.1.

TABLE 13.1
Operational Conditions of Different Dryers

Number	Origin	Drier type	Temp (°C)	Pressure (mbar)	Time (h)
1	Latex	Oven	40	746.6	8
2	Latex	Tray drier	40	746.6	2
3	Latex	Vacuum oven	40	137.06	18
4	Latex	Lyophilisator	−30	0.1	24

13.5.1 Sun Drying

It gives the lowest quality product as there is considerable loss of enzyme activity and the papain can easily turn brown. However, in many countries sun drying is still the most common processing technique for papain. The latex is simply spread on trays and left in the sun to dry.

13.5.2 Hearth Drying

Papain driers can be of simple construction known as hearth. In Sri Lanka and India, they are, generally, simple out-door stoves of 1m height, made out of mud or clay bricks. The latex can also be dried artificially in a home drier in a kind of drying stove constructed by building a chamber of bricks about a meter high, a meter wide and 2 m long. The sides and ends are of bricks with an opening at the end to admit fuel. The top is open. About 30 cm below the top an iron sheet is placed. Upon this 2.5–5 cm of sand is spread to modify and distribute the heat arising from the fire beneath. The coagulated juice is spread upon brown linen stretched upon the frame which is made to fit the top of the drier. Smoking spoils the latex. Therefore, coconut shell or charcoal is recommended as fuel. The fuel is so regulated that drying time, an approximate 4–5 h is affected slowly with temperature preferably below 40°C. Drying is complete when the latex becomes crumble and no longer sticky.

13.5.3 Solar Drying

A multi-rack step solar drier for drying papain has been developed with overall dimensions of 2430 × 1630 × 310 mm 10 trays each of the size of 890 × 460 × 25 mm and two glass layers each of 3 mm. Trays used in the drier are made of aluminium and are provided with semicircular grips so that the same trays can be used for latex collection range of 50–85°C when the ambient temperature is 28–32°C because the butterfly valves provided in the two rear chimneys control the movement of papain production by solar drier as compared to the conventional sun drying method.

13.5.4 Spray Drying

Spray-dried papain has a higher enzyme activity (assaying) than other type of papain, and is totally soluble in water. Extreme care must be taken when handling this form of papain because it can cause allergies and emphysema if inhaled. For this reason, spray dried papain is often encapsulated in a gelatin coat.

13.6 ASSAYING

If papain is to be exploited commercially for an export market or local food industrial uses, it is important to determine the enzyme activity, which is known as assaying. Papain is used for hydrolysation of proteins. Therefore, assays to measure papain activity are based on measuring a product of the hydrolysis. There are two main assay methods: the first relies on the ability of papain to clot milk. It is a low cost but time consuming method. It is also lack of a standardisation to find the clotting point and variations in the milk powder used. In this method, a known amount of papain sample made by dissolving a known weight of papain in a known volume of a solution of acetic acid is added to a fixed amount of milk (dissolving a known weight of milk powder in a known volume of water) which has been warmed to 30°C in a water bath. The contents are thoroughly mixed and then observed until the first signs of clotting are detected. The time taken to reach this stage, from when the papain was added to the milk, is recorded. The experiment is then repeated using different known amounts of papain solutions. The different amounts of papain sample used should give a range of clotting times between 60 and 300 s for optimum results. The activity of the papain sample is then calculated by plotting a graph, finding the time taken to clot milk at an infinite concentration of papain and then using that value in a formula to calculate the activity. To introduce a measure of standardisation, the amount of milk can be fixed at a certain known concentration. This is done by re-acting a known concentration of high grade papain with the milk. The concentration of milk powder solution can then be adjusted to obtain the desired clotting time under fixed reaction conditions. The 'activity of pure papain' at this known amount of milk can then be calculated. Testing the sample papain under the same reaction conditions and known amount of milk will then give an activity relative to the pure papain.

The second method is based on the science of measurement of colour spectrum and absorption of light known as spectrometry. This is the analytical technique for measuring the amount of radiation or colour of light absorbed by a chemical solution. It is known that, for example, a yellow-coloured solution will absorb blue light. Greater is the concentration of yellow colour in the solution, more is the absorption of blue light. This is a useful discovery because certain products of chemical reactions are coloured. More intense colour will be favoured to greater concentration of product. Therefore, by shining the relevant complement colour through the sample liquid the amount of light absorbed can be related to the concentration of product. All colours or radiations of light are not visible to human eyes. The technique used when the 'colours' extend beyond the visible spectrum is known as spectrophotometry and the instrument used is called a spectrophotometer. In the second method to determine the activity of a papain sample, a known amount of papain sample is mixed with a fixed amount of casein (milk protein). The reaction is allowed to proceed for 60 min at 40°C. After this time, the reaction is stopped by the addition of a strong acid. The product of the reaction is known as tyrosine which is known to absorb ultraviolet light (invisible to the human eye). The solutions containing the tyrosine are prepared for analysis using the spectrophotometer. The amount of ultraviolet light absorbed by the solution can be related to the number of tyrosine units produced by the papain sample. Hence, greater the number of tyrosine units, greater is the activity of papain.

13.7 PAPAIN CRYSTALLISATION

A new method of papain crystallizing from fresh papaya latex which gave higher yields than previously reported methods was developed (Monti et al. 2000). This method does not involve the use of sulphydryl reagents. The papain, thus obtained, is practically pure and shows a single band when submitted to electrophoresis on polyacrylamide gel, and is identical to the papain obtained by other methods. In routine enzymatic assays, specific activity was measured using Z-gly-pNP and BAEE as substrates. Papain crystallised by this method, without the use of high concentrations of salts or thiol-containing substances such as cysteine and dithiothreitol, is obtained in the form of a complex with natural inhibitors existent in latex which can be removed by dialysis.

When fresh latex is bubbled with nitrogen for 1 h at room temperature under constant shaking, a volatile substance is released which absorbs light at 250 nm. When present in the latex extract (fraction 2) this substance prevents the spontaneous precipitation of part of the papain that was not identified. A total of 1396 mg papain was obtained from 927 g fresh latex, with a value of 1.51 mg papain per g latex. When fraction 2 was treated by bubbling with nitrogen, 1238 mg papain was obtained from 510 g fresh latex, a total of 2.43 mg papain per gram latex, with a consequent improvement in spontaneous papain crystallisation. Electrophoresis of fresh latex on acid gel (fraction 2) revealed the presence of seven protein components migrating towards the cathode, columns 1, 4 and 6. Two proteins bands were detected in sample of fraction 4 (columns 2 and 3) and electrophoresis of fraction 5 (column 5) showed a single protein band. This protein component had a relative mobility of 0.5 ± 0.03, corresponding to band 3 of the columns 1, 4 and 6, and was identified as papain since papain prepared in laboratory by the classical method (Kimmel and Smith 1954) has the same relative mobility. Gel electrophoresis denaturing of papain obtained by the method described here showed only one band. The pure papain exhibited apparent molecular masses of 21 kDa and the classical papain 21.3 kDa, when G-75 Sephadex was used. The methodology that we used in this work concluded that the papain prepared here has the same properties of the papain obtained from classical method.

13.8 PACKAGING AND STORAGE

The coagulated latex produces about 25% of its weight as dried powder which still contains six to ten percent of moisture. The dried material is mixed with potassium meta-bi-sulphite, ground and diluted with lactose to get concentrate grade papain. A better quality product is obtained if the latex is sieved before drying (Ram 2005). When thoroughly dried, the latex becomes crisp and flaky. It may then be grinded into powder. The powder is packed in air tight bottles or polythene bags. Finally, the whole dried crude papain is then powdered by means of wooden mallet and passed through a 10 mesh sieve. The whole powder is packed in polythene bags, followed by keeping in sealed air tight and light-proof containers (clay pots or metal cans), and finally, kept in a cool and dry place. Metal containers should be lined with polythene. This is not possible at small scale. Crude papain has less value unless it is prepared in the desired grade of demand which is measured by the activity of enzyme. The crude papain, thus prepared, is exported to Europe and United States where it is further refined and sold as powder or in the tablet form under various trade names.

13.9 MEDICINAL, INDUSTRIAL AND PHARMACEUTICAL USES

Papaya can be used as a diuretic (the roots and leaves), anthelmintic (leaves and seed) and to treat bilious conditions (fruit). Parts of the plant are also used to combat dyspepsia and other digestive disorders and a liquid portion has been used to reduce enlarged tonsils. In addition, the juice is used for warts, cancers, tumours, corns and skin defects while the root is said to help tumours of the uterus. Root infusion is also used for syphilis, and the leaf is smoked to relieve asthma attacks. The papaya eating prevents rheumatism and the latex is used for psoriasis, ringworm and the removal of cancerous growth (Nwofia et al. 2012). Papain has been employed to treat ulcers, dissolve membranes in diphtheria, and to reduce swelling, fever and adhesions after surgery. With considerable risk, it has been applied on meat impacted in the gullet. Chemo-papain is, sometimes, injected in the cases of slipped spinal discs or pinched nerves. Precautions should be taken because some individuals are allergic to papain in any form and even to meat tenderised with papain. It is used in meat tenderisers; the major meat proteins responsible for tenderness are the myofibrillar proteins and the connective tissue proteins. Protease enzymes are used to modify these proteins and papain has been extensively used as a common ingredient in the brewery and in the meat and meat processing (Khanna and Panda 2007). Its importance as tenderisers in the food industry is similar to collagenases, which have application in the fur and hide tanning to ensure uniform dying of leather. It has for quite a long time been used in pharmaceutical preparations of diverse food manufacturing applications as the production of high quality kunafa and other popular local sweets and pastries. Papain has been reported to improve meltability and stretchability of Nabulsi cheese with outstanding fibrous structure enhancing superiority in the application in kunafa, pizza and pastries (Abu-Alruz et al. 2009). Also as pharmaceutical products in gel-based proteolytic cysteine enzyme, papain presents antifungal, antibacterial and anti-inflammatory properties (Chukwuemeka and Anthoni 2010). Papain acts as a debris-removing agent, with no harmful effect on sound tissues because of the enzyme's specificity, acting only on the tissues, which lack the Al-antitripsine plasmatic antiprotease that inhibits proteolysis in healthy tissues (Flindt 1979). The mechanism of biochemical removal of caries involves cleavage of polypeptide chains and/or hydrolysis of collagen cross linkages. These cross-linkages give stability to the collagen fibrils, which become weaker and thus more prone to be removed when exposed to the papain gel (Beeley et al. 2000). Papain-based gel has also been reported as potential by useful in biochemical excavation procedures for dentin (Piva et al. 2008). Carpaine, an alkaloid found in papaya leaves, has also been used for medicinal purpose (Sankat and Maharaj 2001). Papain has advantages for being used for chemo-mechanical dental caries removal since it does not interfere in the bond strength of restorative materials to dentin (Lopes et al. 2007). Papain enzyme has a long history of being used to treat sports injuries, other causes of trauma and allergies (Dietrich 1965).

13.10 POTENTIAL AND MARKET OPPORTUNITIES

Papaya, a tropical herbaceous succulent plant which possesses self-supporting stems, grows in all tropical countries and many subtropical regions of the world (Jaime et al. 2007). Moreover, there is no limitation due to seasonality as papaya is available almost round the year. Consequently, there is a need to facilitate the entrepreneurs in understanding the potential of papaya production and the importance of setting up a unit of papain. Yield of papain depends on cultivar, time of taping, nutritional status of plants and region. There are positive correlations between fruit size and papain yield and rainfall and papain yield. Each fruit produces 5–8 g papain depending upon cultivar. A well-managed papaya production has recorded higher papain yield of 8.17 g per fruit and the highest papain of 686.29 g per plant in a period of 6 months (Kumar et al. 2007; Reddy et al. 2012). The variety, CO-2 yielded 100–120 kg papain/acre with an average potential of 200–300 g per tree and a maximum up to 450 g per plant. The yield per hectare is worked out to be 250–375 kg with high yielding papaya varieties. To increase the papain production, high yielding cultivars like, Pusa Majesty, CO-2, CO-5 and CO-6 have been cultivated. Yield of papain increases with the increase in fruit age which is at the maximum between 60 and 70 days. The spacing of 1.6 × 1.6 m and application of 300 g N are found optimum for papain yield in CO-6 at Coimbatore. Application of plant growth regulators like, 2–4D and 2, 4, 5 T @ 10 ppm during blooming period increases the yield of papain. Urea mixed with the plant growth regulators is economical (Ram 2005).

Papain is used in many industries such as, breweries, pharmaceuticals, food, leather, detergents and meat, and fish processing for a variety of processes. Therefore, the end use segments are many in signifying that papain has high export demand. Since there are good prospects for papain market, the papaya production and extraction of papain can be a high source of income even for small farmers. Lanced fruits are sold in market after ripening because latex extraction does not impair its taste or other qualities in any way. Remaining fully grown but green and scarred fruit can be used for processing after extraction of latex for candy or confectionary (petha) making. Petha production is a lucrative industry and will be a great success for generating extra income from these fruits.

14 Fruiting, Harvesting and Post-Harvest

Papaya flowers are borne on inflorescences that appear in the axils of the leaves. Female flowers are held close against the stem as single flowers or in clusters of 2–3 (Chay-Prove et al. 2000). Male flowers are smaller, numerous and are borne on 60–90 cm long pendulous inflorescences (Nakasone and Paull 1998). Bisexual flowers are intermediate between the two unisexual forms (Nakasone and Paull 1998). Male flowers have no ovary and do not produce a fruit. They contain stamens bearing pollen that can pollinate a papaya flower with an ovary, making it to produce a fruit. It is known as polycarpic fruit crop. Flowering and fruiting continue throughout lifespan of plant but economic life is only for 1–3 years. Fruit shows a double sigmoid type growth pattern during development. A range of variation of 4–6 months for first flowering date, 8.5–11 months to first harvest and 1–7 kg for fruit weight were recorded (Ocampo et al. 2006).

In modern cultivars, one papaya leaf can sustain the development of three to four fruits. However, there are indications of poor adjustment capacity of source to sink ratios in fruiting papaya plants, presumably because the fruits have low capacity to attract assimilates (Zhou et al. 2000). In most of the crops, biomass allocation to the harvested organ is the yield component, and most susceptible to selection and breeding (Bugbee and Monje 1992).

Soil type, mulching, irrigation and fertilisation influence the water and nutrient supply to the plant, which in turn affect the nutritional quality of fruit harvested. The effects of mineral and elemental uptake from fertilisers by plants are variable. High calcium uptake by fruit has been reported to reduce respiration rates and ethylene production delays ripening, increases firmness, reduces incidence of physiological disorders, decays and ultimately increases shelf-life. High nitrogen content is often associated with reduced shelf-life and decay (Kader and Rolle 2004). Over feeding of nitrogenous fertilisers results in softening of papaya fruit (Desai and Wagh 1995).

There is a progressive increase in total sugars, vitamins (A and C), minerals (P, K, Ca), xanthophylls and carotene pigments as the fruit matures. The nutritional composition of the fruit at harvest varies widely depending on cultivar, maturity, climate, soil type and fertility. Ascorbic acid and carotenoid contents increase with maturation and ripening of papaya fruit (Lee and Kader 2000). Also, ascorbic acid levels in fruit are influenced by the availability of light to the crop and to individual fruits. In general, the lower the light intensity, the lower the ascorbic acid content of plant tissues; best quality of papaya fruit is determined largely by sugar content, development under full sunlight in the final 4–5 days to full ripening on the tree (Samson 1986). Lower temperature (<10°C) decreases fruit growth, sweetness and fruit size of papaya (Desai and Wagh 1995). Rainfall may influence the composition of the harvested plant part and its susceptibility to mechanical damage and decay during subsequent harvesting and handling operations (Sankat and Maharaj 2001). Papaya responds well to adequate irrigation, which helps rapid fruit development and regular fruit yield. Pre-harvest application of pesticides and growth regulators does not directly influence fruit composition but may indirectly affect it due to delayed or accelerated fruit maturity (Nakasone and Paull 1999).

14.1 MATURITY, RIPENING AND HARVESTING

The main problem faced in papaya fruit marketing is identification of optimum harvest maturity to ensure adequate fruit ripening to good eating quality (Workneh et al. 2012). Fruit quality and storage behaviour are influenced greatly by maturity of fruit at harvest. Immature fruits are more subject to shrivelling and mechanical damage, and are of inferior flavour quality when ripe. Overripe fruits are likely to become soft and mealy, with insipid flavour soon after harvest. Fruits picked either too early or too late are more susceptible to post-harvest physiological disorders than fruits picked at the proper maturity. Most of the fruits reach their best eating quality when allowed to ripen on the plant. However, some fruits are usually picked mature but unripe so that they can withstand the post-harvest handling system when shipped long distance (Kader and Płocharski 1997).

Fruits require 125–140 days from flowering to maturity. Change in fruit skin colour has been used as a harvest index criterion to judge maturity. Portion of fruit exposed to sunlight becomes dark yellow in colour. Papaya fruits should be harvested when the colour of the skin changes from dark green to light green and when one yellow streak begins or some skin yellowing development from the base upwards takes place (Figure 14.1). However, the yellow colouration pattern is not necessarily restricted to longitudinal stripes, and yellow coloured sites can appear almost anywhere on the fruit skin (Peleg and GomezBrito 1975). Fruits in this condition will continue to ripen normally after harvest. The papaya flavour is at its peak when the skin is 80% coloured. Flesh colour changes from green to yellow or red depending upon cultivar (Kader 2006). For the local market, in winter months, papayas may be allowed to colour fairly well before picking, but for local market in summer and

FIGURE 14.1 **(See colour insert.)** Papaya fruits with the sign of ripening: (a) Pusa Majesty and (b) Pusa Dwarf.

for shipment, only the first indication of yellow is permissible. Technically fruits are harvested at colour break of 1/4 yellow for export or at 1/2–3/4 yellow for local markets, depending on cultivars, ripening characteristics and season. Those fruits harvested before this stage could fail to show complete ripening, while those harvested later could be more susceptible to damage and bruising during handling (Paull 1993a,b). Less mature fruits are lower in sugar and ripen poorly (Kader 2006). Fruit ripening in papaya cultivars varies widely in terms of softening, skin colour changes and shelf-life (Zhang and Paull 1990; Thumdee et al. 2007). At fruit maturity, latex of fruits becomes watery and T.S.S at harvest should be minimum 8° Brix. Hawaii specifies a minimum total soluble solid of 11.5% and fruit showing at least 6% surface colouration at the blossom end region with proper fruit size (Quinta and Paull 1993). The fruit is size-graded, treated with hot vapours of ethylene bromide to kill fruit fly, wrapped in newspapers and packed in boxes before shipping. Such wrapped fruits will ripen and turn completely yellow in colour at room temperature within 2–4 days. The fruit should be harvested with a 20-mm portion of stem attached and must be packed carefully in a single layer in a carry box with the stem-end resting on the bottom. The bottom of the carry box should be covered with paper wool or other noncontaminant material, which will absorb latex. Latex should, therefore, be allowed to drain from the stem-end onto the paper wool where it will be absorbed. Care should be taken not to have any latex dripping onto the fruit during harvesting (DAIS 2009). The papayas must be packed in a way that ensures they are sufficiently protected for export purposes. Packing material used inside the carton must be new, clean and must be shaped in such a manner that it cannot cause any damage to either the inside or outside of the fruit. The usage of materials, such as papers and stickers with company details on them, is permitted provided that no toxic inks, dyes, or glues have been used. The packaging must be free of all other materials.

Packaging provides protection from physical damage during storage, transportation and marketing (Irtwange 2006). Azene et al. (2011) reported on the effects of different packaging materials and storage environment on post-harvest quality of papaya fruit. There are variety of packages, packaging materials and inserts available. Nowadays, produce is transported and sold in an enormous way of packages constructed of wood, fibre board, jute, or plastics. An important supplement to proper temperature and relative humidity management is the use of modified atmosphere (MA) (Azene et al. 2011). Modified atmosphere packaging (MAP) of fresh produce relies on modification of the atmosphere inside the package, achieved by the natural interplay between two processes, the respiration of the product and transfer of gases through the package, which leads to an atmosphere richer in CO_2 and poorer in O_2. This atmosphere can potentially reduce the respiration rate, ethylene sensitivity and production, decay and oxidation and hence delays ripening and senescence (Kader and Rolle 2004). MAP also relates to packages and film box liners with specific properties that offer a measure of control over the composition of the atmosphere around the produce (Irtwange 2006). The principal plastic materials for MAP that can be used with fruits and vegetables include polybutylene, low-density polyethylene (LDPE), high-density polyethylene (HDPE), PP, PVC, polystyrene, ionomer, pliofilm and polyvinylidine chloride (Schlimme and Rooney 1994). The permeability of films to gases (including water vapour) varies with the type of material from

TABLE 14.1
Permeability Characteristics of Some Plastic Films with Potential for Use as MAP of Fresh and Lightly Processed Produce

	Transmission Rate		
Film Type	O_2[1]	CO_2[a]	H_2O Vapour[b]
Low-density polyethylene (LDPE)	3900–13,000	7700–77,000	6–23.2
Medium-density polyethylene (MDPE)	2600–8293	7700–38,750	8–15
High-density polyethylene (HDPE)	52–4000	3900–10,000	4–10
Polypropylene (PP)	1300–6400	7700–21,000	4–10.8
Polyvinylchloride (PVC)	620–2248	42638.138	>8

Source: Adapted from Schlimme, D. V, and M. L. Rooney, 1994, in *Minimally Processed Refrigerated Fruits and Vegetables* (ed. R. C. Willey), New York: Chapman and Hall, 135–179.

[a] Expressed in terms of $cm^3m^{-2}day^{-1}$ at 1 atm.

[b] Expressed in terms of gm^{-2} day^{-1} at 37.8°C and 90% relative humidity.

which they are made, temperature, in some cases humidity and the accumulation and concentration of the gas and the thickness of the material (Thompson 2001). The characteristics of the main types of plastic films with their potential uses in MAP are summarised in Table 14.1.

Each carton must display the regulations with following details in unbroken, legible, permanent letters visible from the outside as described (Naturlande 2000).

Identification	Type of Product 'Papaya'	Origin of the Product	Commercial Characteristics
Name and address of the exporter and packer	When the contents are not visible, mention the name of the variety	Country of origin, and optionally, national, regional or local description	Class, size (reference letter or weight class), number of fruits (optional) and net weight (optional)

In India, fruit harvesting is done by picking individual fruits by hand and not allowed to fall on the ground. In the peak season, the fruit should be harvested about thrice a week. During the rest of the season, the physiologically mature fruit should be identified through regular scouting. Home growers may twist the fruit to break the stem, but in commercial operations, it is preferable to uses a sharp knife to cut the stem and trim it to level with the base of the fruit (Ram 2005). However, to expedite harvesting of high fruits, growers furnish their pickers with a bamboo pole with a rubber suction cup at the tip. With the cup held against the lower end of the fruit, the pole is thrust upward to snap the stem and the falling fruit is caught by hand. One man can thus gather 3.5–4.5q daily. It is harvested by twisting the fruit on Pusa Dwarf and Pusa Nanha trees; the operation can be done without mechanical aids. Ladders are often used to harvest tall trees. It has been calculated that manual picking and field sorting constitute 40% of labour cost of the crop. The mechanical aid was tested and results indicated that a machine with one operator and two pickers could harvest 450 kg of fruit per hour, equivalent of eight men hand-picking. The latex oozing from the stem may irritate the skin and workers should be required to wear gloves and protective clothing (DAIS 2009).

14.2 YIELD

The yield varies widely according to variety, soil, climate and management of the orchard. About 20–50 fruits are obtained from a plant and on an average fruit yields vary from 50 to 75 tonnes per hectare in a season from a papaya orchard depending on spacing and cultural practices during the first year and it reduces to be 20–25 tonnes per hectare in the second year. Maximum average fruit weight (1.20–3.05 kg), length of fruit (22.45 cm), width of fruit (33.35 cm) and fruit yield (120.11 tonnes/ha) were recorded when planting with a spacing of 2. 5 × 2.5 m in papaya cv. Coorg Honey Dew was done (Singh et al. 2010). Fruits are stored in a single layer in bamboo baskets and covered with a thin layer of paddy straw and sent to local markets, while for distant marketing, individual fruits are wrapped in newspaper and packed in single-layer bamboo baskets and after putting saw dust as a lining material and covered with a layer of paddy straw (Naturlande 2000).

14.3 POST-HARVEST MANAGEMENT

Mature papaya fruits ripe within 2–4 days after harvesting. Paull (1993) reported that ripening ranges from 7 to 16 days from the colour break stage. The rate of softening could differ between cultivars with respect to the rate of respiration, ethylene production, skin degreening and flesh colour development (Paull 1993). Shelf-life of fruit can be affected by several factors, namely, respiration, biological structure, ethylene production and sensitivity, transpiration, developmental processes and physiological breakdown (Irtwange 2006). It can be stored under controlled atmospheric conditions, namely, 2% oxygen, 5% carbon dioxide and a temperature of 16°C can extend shelf-life up to 14 days. Fruits stored >20°C will be affected by fungal diseases (*Colletotrichum* sp.) and <10°C by chilling injuries. Fruits can be held at 30°C and high atmospheric humidity for 48 h to enhance colouring before packing. Standard decay control has been a 20-min submersion in water at 49°C, followed by a cool rinse. In India, dipping in 1000 ppm of aureofungin has been shown to be effective in controlling post-harvest rots. An aqueous solution of carnauba wax and thiabendazole over harvested fruits gives good protection from post-harvest diseases. Papayas must be treated before export to avoid introduction of fruit flies. Fruits picked 1/4 ripe are pre-warmed in water at 43°C for about 40 min, and then quickly immersed for 20 min at 48°C. This double-dipping maybe replaced by irradiation. Fruits that have had hot water treatment, followed by irradiation at 75–100 krad and storage at 3% oxygen and 16°C for 6 days can be expected to have a market shelf-life of 8 days, while gamma irradiation (25–50 krads) delayed ripening up to 7 days. Fruits treated at 100 krads also slightly accelerated ripening in storage. The carotenoid content was unaffected but ascorbic acid was slightly reduced at all exposures. Even the lowest level of irradiation inhibited fungal growth. Partly ripe papaya stored below 10°C will never fully ripen. This is the lowest temperature at which ripe papayas can be held without chilling injury.

The degree of maturity will indicate whether papaya fruit should be exposed to ethylene. Papayas that are fully mature at harvest should not be ripened with ethylene if they are to be stored for an extended period of time. Papayas of minimum commercial maturity will benefit from a treatment of ethylene with an improvement in texture and colour. A standard banana ripening room may be used for papayas as well. If not using pressurised ripening rooms, then air stack the boxes (at least 2″ between boxes) to ensure proper air circulation. Leave 1 1/2 feet between walls and pallets and about 6″ between pallets. Depending upon desired shipping time, bring the pulp temperature to the range of 20–25°C and apply 100 ppm ethylene for 24–48 h. The green fruits are ripened successfully by 6–7 days treatment with ethylene gas in airtight chambers at 25°C and 85%–95% humidity, followed by hot water treatment. The actual time of exposure to ethylene depends upon maturity of the fruit. The yellowing of fruit indicates that papayas are producing ethylene and the generator is no longer needed. Once the desired level of ripeness is attained, reduce the temperature up to 10°C for partially ripe papaya or 7°C for ripe fruits.

Papaya is a climacteric fruit with characteristic respiratory peak and ethylene production pattern during fruit ripening. Respiration rates alter during a natural process of fruit ripening, maturity and senescence (Desai and Wagh 1995; Irtwange 2006). Papaya experiences a marked and transient increase in respiration during its ripening, which is associated with increased production of and sensitivity to ethylene (Desai and Wagh 1995). The sudden upsurge in respiration is called the "climacteric rise", which is considered to be the turning point in the life of the fruit. After this senescence and onset of fruit deterioration, reduced food value for consumers and increased loss of flavour and salable fruit weight results. To extend the post-harvest life of climacteric fruits, their respiration rate should be reduced as far as possible (Irtwange 2006).

At the onset of ripening, respiration rises to a maximum (climacteric peak) and ethylene production increases with a similar pattern (Bron and Jacomino 2006). One of the effects of storage under MAP is for levels of ethylene produced by the fruit to diminish, along with changes in colour and texture, while changes in sugars and acids responsible for some of the flavour proceed normally (Wills et al. 1989). Lazan et al. (1990) reported that there was a concomitant decrease in internal ethylene concentration of papaya fruit packaged with polyethylene film, which may be instrumental in delaying ripening of the sealed fruit. The rate of ethylene production of papaya fruit stored at 20°C ranges from 10 to 100 $\mu l\ kg^{-1}\ h^{-1}$ (Nakasone and Paull 1999). The ethylene forming enzyme activity was found to be maximum in the exocarp of three-quarter-ripe fruit (Sankat and Maharaj 2001). The level of 1-amino-1-cyclopropane carboxylic acid (ACC), the substrate for ethylene forming enzyme, is initially low in fruit mesocarp tissue during ripening, increasing threefold when the peak of ethylene synthesis occurs. Ethylene-treated papayas ripened faster and more uniformly in terms of skin de-greening, softening and flesh colour. The enzymes, polygalacturonase (PG), pectin methylesterase (PME), xylanase and cellulose were reported. There is a relationship between PG and xylanase and fruit softening (Paull et al. 1999). The peak in xylanase and PG activity occurs when the fruit has 40%–60% skin yellowing (Paull 1993). Since papaya ripens from the inside outwards, the effect of ethylene treatment is to accelerate the rate of ripening of the mesocarp tissue nearer the skin that has not started to soften. The already well-softened mesocarp that is near to the seed cavity is not responsive to ethylene. Ethylene is not recommended commercially, as the rapid softening severely limits marketing time (Paull et al. 1997). Thus, fruits can be kept in pre-climacteric by controlling atmospheric gas composition in special storage or modifying the atmosphere with in a package. Packaging that absorbs ethylene, carbon dioxide, or oxygen is being developed to control or retard the ripening process (Desai and Wagh 1995). The post-harvest life of fruit can, thus, be extended through low-oxygen concentration and slightly high carbon dioxide level and decreasing the storage temperature. The optimum gas composition is the range of oxygen and carbon dioxide level that would minimise physiological disorder, and reduce respiration rate and ethylene production during storage (Kays 1997). A decrease in the rate of respiration increases the shelf-life of fruits (Wills et al. 1989). The removal of oxygen from the storage environment is another important effect of respiration. As a consequence, the rate of respiration is important for determining the amount of ventilation required in the storage area. This is also critical in determining the type and design of packaging material that can be used, as well as the use of artificial surface coatings on the product (Kays 1997).

Harvested fruits continue to respire and lose water even when they are attached to the parent plant, only difference being that losses are not replaced in the post-harvest environment. Water loss through transpiration is the first stage for loss of marketable weight and textural quality, softening, loss of crispness and juiciness, followed by reduction in nutritional quality (Wills et al. 1989). Papaya belongs to fruits with high-moisture loss rate (Nakasone and Paull 1999). Gradient in water deficit, which is a parallel gradient in tissue softness, occurs in the mesocarp tissue of papaya fruit (Lazan et al. 1990). At high relative humidity, produce maintains salable weight, appearance, nutritional quality and flavour, while wilting, softening and juiciness are reduced (Nakasone and Paull 1999). Packaging of fruit with polyethylene film retards development of water stress and softness in the fruit tissues.

14.4 PROCESSING

Papaya fruits are very sensitive to pressure. The fungus *Colletotrichum gloeosporioides* can easily spread through cuts or where the fruit is attached to the stem. The fungicide baths used in conventional papaya plantations are not permitted. Generally, fruits between 220 and 600 g are in demand on the international markets. It is recommended to harvest the fruits for export as soon as the tips turn yellow from green. The fruit pulp can attain a Brix value of 10%–11.5%. After harvesting, the fruits are washed in hot water, which varies in temperature and duration according to the type of fruit. On average, the papayas are bathed at 49°C for 20 min or at 42°C for 30 min. After this treatment, the fruits are slowly cooled down to room temperature, dried, sorted, classified, packed and then stored in the cool place until being shipped. The papayas must be fresh, healthy, clean, well-developed, ripe and free from bruises, frost-damage and strange taste or smell with different classes (Naturlande 2000).

A. Class extra: papayas in this class must be of the highest quality. They must possess the characteristics typical of their variety and/or trading type. The fruits must be unblemished, with the exception of very light surface flaws that do not detract from the fruit's general appearance, quality, the time it will keep and its presentation.

B. Class I: papayas in this class must be of good quality. They must possess the characteristics typical of their variety and/or trading type. The slightly misshapen, light flaws in the skin caused by friction or by other means provided the area does not exceed 3% of the total surface area of the fruit. These blemishes are permissible, provided they do not detract from the fruit's general appearance, quality, the time it will keep and the presentation of the bunch or cluster in their packaging.

C. Class II: this class is composed of those papayas that cannot be placed in the upper classes, yet which fulfil the definitions of minimum requirements. The faults for shape defects, colour defects, skin flaws caused by scratches, friction, or other means provided that <10% of the total surface is affected are allowed and that the papayas retain their essential characteristics in terms of quality, preservation quality and presentation. The flaws are not permitted to affect the fruit's pulp.

14.4.1 Size Classification

The papayas are sorted according to the weight. The fruits must weigh at least 200 g and graded as (A) 200–700 g, (B) 700–1300 g, (C) 1300–1700 g, (D) 1700–2300 g and (E) >2300 g by international standards.

A variety of products such as jam, jelly, nectars, ice-cream, sherbet, yogurt, fruit leather and dried slices may also be made from the ripe fruit. Unripe papaya makes a good concoction of vegetable stew, salad, or pickle (Workneh et al. 2012).

14.4.2 Dried Papayas

Dehydration is the oldest method of making food storable for longer periods. It is based on the fact that micro-organisms tend to cease growing below a certain level of water content. During drying, it is important to extract the water from the fruit as carefully as possible. The most important features are a good circulation of air and not too high temperatures (Momenzadeh et al. 2010).

After harvesting, the fruits are sorted as only fresh and unripe. Fermented fruits cannot be used for drying. Papayas must be washed very carefully in order not to damage them. Afterwards, inedible parts such as leaves, seeds, pips, heartwood and skins are removed. The fruits are now cut into pieces of equal size and laid out to dry in the air and sun in thin layers on racks, in solar dryers (drying tunnels), drying ovens or microwave (MW) drying (artificial drying at 60°C). The MW drying process occurs between 40 and 50°C temperatures and best fitting traits to predict the drying characteristics of papaya slices during the drying process at 60°C (Yousefi et al. 2013).

Before they are packed, the fruits are inspected and sorted again, to rid them of discoloured, skin remnants, seeds, and so on. The packaged fruits can now be labelled and stored prior to being shipped. During and after drying, the dried fruits are not permitted to be treated with methylbromide, ethylene oxide, sulphur oxide, or with ionising radiation. The dried fruits can be packed in consumer packs, or wholesaler packs (bulk) in bags made of sealable foils, impermeable to steam (e.g., polyethylene or polypropylene). Before sealing, a gas (e.g., nitrogen) may be added (nitrogen flushing). The dried fruits should be stored in dark areas at low temperatures and relative humidity. Under optimum conditions, dried fruits can be stored for up to 1 year.

14.4.3 Marmalades

Jams are basically preparations made of fruit and various sugars that are made conservable mainly by heat treatment. The half-set yet spreadable consistency of these products is achieved by releasing the pectin found in the fruit pulp during the boiling process, and using this together with further pectin added to form a jelly-like mass. After harvesting, the fruits are sorted, because only those that are fresh, ripe and not rotten can be used to make jams. Jams can also be made from previously prepared, frozen fruits and pulp (Saran and Choudhary 2013). The fruit should be washed very carefully as it can easily be damaged. Peeling is often done manually, with knives, yet sometimes, the skin is loosened with steam and then subsequently rubbed away mechanically. Finally, the fruits are sorted again to remove any blackened pieces, bits of peeling, seeds, and so on. The peeled fruits are then pulped and sugar added. They might also be mixed with water or fruit juice. Boil 1.0 kg pulp of ripe firm peeled fruit with 100.0 mL water and 3.0 g citric acid. The mixture is heated to 70–80°C and boiled down, with consistent stirring, to 65°C until shortly before it reaches the desired consistency. Add 750.0 g sugar and cook up to thick consistency. The end point is confirmed by the sheet test. Boiling mass is allowed to fall after cooking from a serving spoon, which will flow in the form of a sheet. If necessary or desired, citric acid, pectin spices and natural flavourings can be added, and the mixture again briefly heated to 80°C (Medina et al. 2003). The liquid mass is now poured into jars, vacuum-sealed and pasteurised. After the heating process, the jams are first cooled to 40°C, and then subsequently down to storage temperature, labelled and finally stored. The jams should be stored in a dark, cool room at temperatures of a maximum of 15°C. Under optimum conditions, jam may be stored for 1–2 years.

14.4.4 Canned Papayas

Canned foods are products that can be stored over a long period in airtight containers (metal or glass jars). They are preserved mainly by heat treatment, during which the micro-organisms present in the fruit are significantly reduced in number, or their development so restricted, that they are prevented from spoiling the product. During the process involved in turning fresh fruit into canned products the fruits are sorted, because only those that are fresh, ripe and not rotten can be used to make jams. The fruit should be washed very carefully as it can easily be damaged. This follows the procedure of removing leaves, wooden pieces, pips or seeds and peel. Peeling is often done manually, or with knives, yet sometimes, the skin is loosened with steam and then subsequently rubbed away mechanically. Finally, the fruits are sorted again to remove any blackened pieces, bits of peeling, seeds, and so on. The peeled fruit can be cut into a variety of shapes, according to type. The shape of the cut fruit must be given on the can (slices diced, pieces, etc.). The cut pieces are now filled into jars or cans and covered with syrup. Additional information must be given on the can such as the sugar content of the syrup. Sugar concentration of the syrup is described on the can, namely, very lightly sugared (9%–14%), lightly sugared (14%–17%), sugared (17%–20%) and strongly sugared (>20%), vacuum sealed and pasteurised or sterilised. After the jars or cans have been vacuum sealed, they are either pasteurised (temperatures >80°C) or sterilised (temperatures >100°C). After the heating process, the canned fruits are first cooled to 40°C and then subsequently down to storage temperature.

14.4.5 Jelly

Mix 1.0 kg grated pulp of fully mature peeled, but somewhat, raw fruits and 1.0 kg ripe papaya pulp with 2.5 L of water and 10.0 g citric acid. Boil for 30 min, cool and allow settling for 2 h. Separate the supernatant and filter. Formation of single clot with small quantity of ethyl alcohol added to test samples indicates high pectin content. Concentrate further, if necessary, to obtain single clot. Cook gently the extract with equal quantity of sugar to obtain the end point indicated by the formation of sheet. For packing of hot jelly, cover with a layer of melted wax and close the lid for safe storage (Medina et al. 2003).

14.4.6 Fruit Pulp

Juice found in the peel should not be allowed to mix with pulp. The multitudinous and highly active enzymes should be inactivated during processing. All machine parts that come into contact with the pulp must be constructed out of stainless steel to prevent discolouring. In order to make papaya marmalade, only fresh, ripe and not mouldy fruits should be used. After harvesting, the fruits are then sorted and washed in a water bath at 50°C for 20 min. Next, they are sent through a steam tunnel, where they are treated with steam (100°C) in order to prevent latex from oozing out of the peel; to deactivate the enzymes in the peel; to clean the surface of the fruit; to reduce the number of micro-organisms, and to soften the outer parts of the fruit in order to increase the amount of pulp produced. The fruit is then cooled down with water jets for 3–4 min, subsequently peeled, and then fed into a strainer (0.6 mm screen) to remove the kernels. Afterwards, the pulp is generally treated with a 50% citric acid solution in order to lower the pH value to 3.0–3.5. Another strainer (<0.5 mm screen) removes the fine fibres and particles, before the pulp is heated to 93.0–96.0°C in a heat exchanger for 2 min. The papaya pulp can now be filled (up to 5.0 kg) into tin (lead-free) cans whilst still hot, whereby the cans are sealed while being steamed, the temperature is maintained for 5 min, and then rapidly cooled down. At temperatures of around 15°C, the pulp can be stored for up to 1 year. After pasteurising, the pulp can also be cooled down and filled into polyethylene bags placed in 50–200 kg barrels. It is then rapidly frozen and can be stored at –18°C for 18 months. Pulp that has been filled under antiseptic conditions (bag-in-box) can be stored for up to 1 year at room temperature. The pulp/juices can be packed into single or wholesale packages (bulk) consisting of glass jars, tin cans, or polyethylene or polypropylene bags, and also filled antiseptically into 'bag-in-boxes'.

14.4.7 Vegetable and Raita

Wash, peel, cut in two pieces, remove seeds, slice, rewash and cook as other vegetables. For large-scale consumption of early fallen green fruits, supply them for use as vegetable and raita. Boil grated pulp of peeled green papaya with equal amount of water. Press to squeeze out water. Mix curd with ladle, add grated boiled pieces and mix. Mix edible salt and spices (cumin, black pepper, coriander, etc.) according to taste and market (Medina et al. 2003).

14.4.8 Pectin

Unripe green fruits or fruits after obtaining latex (unsuitable for product making) may be used for pectin extraction. Green fruit is rich in pectin (10%) on dry weight basis. Pectin has extensive applications in food and medicinal industries. Pectin can also be extracted from peel waste of green fruit (Medina et al. 2003).

14.4.9 Pickle

Blanch slices of peeled green fruit in boiling water for 3 min. Drain and sprinkle 100.0 g salt and dry under shade. Mix spices, namely, 10.0 g each of powdered red chillies, cardamom, large cumin and black pepper for 1 kg slices. Fill in jars and cover with vinegar for curing (2–3 weeks) and storage with airtight lids (Medina et al. 2003).

14.4.10 Chutney

Cook 1.0 kg pulp of ripe, firm, peeled and grated papaya with chopped onion (100.0 g) and 50.0 g salt till it gets soft. Then add to it 50.0 g ginger, 15.0 g garlic after chopping and 10.0 g each of the powdered aniseed, red chillies, cumin, large cardamon, cinnamon, black pepper and 2.0 g headless cloves. Cook gently to desired consistency. Add sugar (3/4 kg). Cook again to thick consistency. Add vinegar (200 mL) and cook for 5 min. Fill hot into the cleaned dry jars, followed by capping (Medina et al. 2003).

14.4.11 Sauce

Concentrate 1.0 kg strained pulp containing 20.0 g sugar up to 1/3 of its original volume in the presence of suspended spice bag containing 50.0 g chopped onion, 5.0 g garlic, 50.0 g ginger, 10.0 g powdered spices and 5.0 g red chillies. Press out spice bag occasionally and squeeze it out finally to obtain maximum spice extract. Add 15.0 g salt and remaining 40.0 g sugar and cook to thick consistency. Add

450 ml vinegar and cook again to the end point. Add and mix preservative after dissolving in minimum quantity of water. Heat up to boiling and fill in hot pack (Medina et al. 2003).

14.4.12 Burfi

Boil the grated pulp of peeled green papaya for 5 min and press out excess water. Fry in equal amount of milk fat. Add sugar equal to the grated material, mix properly and heat for 2 min. This material is spread in the form of 1–2 cm thick uniform layer on aluminium/steel tray smeared with fat and is allowed for drying. Cut into pieces of suitable size as per the need of consumers (Medina et al. 2003).

14.4.13 Toffee

Cut slices of suitable size and concentrate 1.0 kg sieved pulp to 1/3 volume and cook with added sugar (600.0 g), glucose (100.0 g) and hydrogenated fat (100.0 g) till a speck of the product put into water forms compact solid mass/lump. Make thick paste of 100.0 g skim milk powder in minimum quantity of water and mix with the boiling mass. Spread a 1–2 cm thick layer of the cooked mass over/trays smeared with fat. Add flavouring material as per requirement, cool, cut and wrap in butter paper (Medina et al. 2003).

14.4.14 Leather

Mix thoroughly powdered sugar (50.0–75.0 g), citric acid (0.5 g) and KMS (3.0 g). Smear steel tray with fat and spread the above mixture in 1.0 cm thick layer. Dry in a home drier at 55–60°C. Roll the dry leathery product for storage as slab or cut it into pieces of suitable size (Medina et al. 2003).

15 Marketing and Economics

Small and commercial farmers in many countries grow papaya for both local and foreign markets. The local markets prefer medium and large-fruited varieties that have red and yellow flesh. Papaya fruits for export are usually small or of medium size (Codex 2005; Stice et al. 2010) with red, orange or yellow flesh (Pesante 2003; Picha 2006). Both hermaphrodite fruits (oblong to pear-shaped) and female fruits (roundish) are usually accepted by consumers in some countries, but only those fruits are preferred that are fresh, free from bruises, blemishes and uniform in size and ripening. The latest Codex alimentation standards for papaya have been amended in 2005 and included standards regarding quality, size, uniformity, packaging, labelling, contaminants and hygiene (Codex 2005).

15.1 MARKETING

Markets of papaya for domestic use, export, processing plants etc. are available in India. There are several fruit traders and consolidators who buy papaya fruits from producers and transport to the different cities for retailers. Some big fruit processors and exporters directly enter into the growers market through contract farming. Other growers also have their own market outlets. Prices in the market depend upon supply and demand situation, and vary according to fruit quality, variety, seasons, etc. However, papaya has a big export demand but only a few exporting companies have access to markets because of the stringent quality requirements and big capital outlay. Ripening, packing, transportation and selling are the main marketing functions involved in the process of papaya marketing. Fruits are packed with the help of paper in corrugated fiber board (CFB). Better packing always helps in maintaining the quality and also reducing the losses during transit on the account of spoilage. Generally, the papaya fruits are being transported by trucks. The producer sells its maximum quantity through commission agent-cum-wholesaler at the farm level. About 76% of the sample farmers sold their produce to commission agents-cum-wholesalers and remaining 24% of the farmers sold them to the distant markets (Devi and Saran 2014). The commission agent in the distant market arranged for sale in the market and charged a commission (10%) for sale of proceeds from producer and seller. Two marketing channels have been identified by Shivannavar (2005) as follows:

Channel-I: Producer - commission agent-cum-wholesaler - retailer - consumer.
Channel-II: Producer - commission agent - wholesaler - retailer - consumer.

In India, Channel-I was most commonly used (76%) by the sample papaya growers. Here the commission agent-cum-wholesaler himself comes to the farm and fixes the price and then takes papaya fruits to market. Most of the growers prefer Channel-I for low marketing cost incurred in this channel and there was no risk of price fluctuation after taking the produce to the market. In Channel II, the farmer himself has taken the produce to distant market where he has to bear all marketing costs such as transportation, unloading, weighment and commission charges. But no doubt, the producer will get a better price for his produce in the market than at the farm level. Shivannavar (2005) reported that all the papaya-growing sample farmers expressed the severity of virus attack along with labour intensiveness and water scarcity during papaya cultivation. The other problems were: lack of technical know-how (79%), irregular power supply (78%), higher initial investment (68%), smaller holdings (37%) and duplication of seeds (20%). It could be seen that all the respondents opined that markets are far away from the farm, over 82% of the respondents opined that higher commission charge was also another major problem in marketing of papaya. The other problems were: lack of availability of market information (79%), storage problem (76%), price fluctuations (37%) and lack of skilled labour for packing (19%).

15.2 ECONOMICS

Papaya is a profitable crop and provides an income next to bananas. It can be grown for fruits, seed and for papain extraction. In a properly managed papaya orchard, crop yield is very high and gives high amount of profit. Demand for papaya in the United States has been growing due to many factors, including more awareness of the fruit's health benefits and Asian and Hispanic population growth. Tropical fruit growers in South Florida are in search of profitable alternatives to increase revenue and to ensure that their operations remain profitable. While there appears to be an opportunity for these growers to take advantage of this growing market, given their closer proximity to the market and in light of recent restrictions being placed on papaya imported from Mexico (the number one US papaya supplier), the analysis suggests that they are less likely to do so because of the relatively unattractive returns associated with producing the crop under current conditions. The average grower would invest $11,322 per acre for a net return of only $278 per acre over the first 2-year period. Even with a singular focus on gross margin only (not advisable), the return would be $1880 per acre or $940 per acre per year (Table 15.1). Since not much can be done about increasing growers' profits, growers would need to increase their output for this to become an attractive alternative. A 10% increase in output at current prices, for example, would cause net profit to increase by more than 300%. At present, growers are severely constrained by the widespread presence of PRSV, which has severely reduced the output and has limited the production cycle to only 2 years, thus, increasing the unit cost of production. Preliminary results indicate that with modest increases in production costs and growing PRSV-resistant GM varieties, growers could increase their output three- to four-fold and extend the harvesting season by another year to compete and be profitable (Evans et al. 2012).

Generally, Indian farmers can get Rs. 35,000–40,000 from one hectare of papaya plantation for fruits. Success stories of two farmers, one form Sholapur district of Maharashtra and another from Jalpaiguri district of West Bengal have been reported. Maharashtra farmers obtained a yield of 138.75 tonnes/ha with a net profit

TABLE 15.1
Cost per Acre of Establishing and Producing Papaya on a South Florida Five-Acre Orchard

	Unit	Year 1 (Establishment)	Year 2 (Full Production)	Year 1 and 2 Total
Estimated yield	Lb/acre	6500.00	22,500.00	29,000.00
Estimated price	$/lb	0.40	0.40	0.40
Total receipts	dollars	2600.00	9000.00	11,600.00
COSTS				
Pre-harvest costs	$/acre			
Soil preparation (disc and bed)		100.00		100.00
Trees		338.00		338.00
Planting and other costs		225.00		225.00
Irrigation		250.00	250.00	500.00
Fertilisers		663.00	866.00	1529.00
Herbicides		100.00	200.00	300.00
Insecticides		356.00	469.00	825.00
Fungicides		404.00	584.00	988.00
Mowing		120.00	120.00	240.00
Labour (weeding, fertilising, etc.)		684.00	812.00	1496.00
Repairs		40.00	41.00	81.00
Interest on pre-harvest costs		98.00	100.00	198.00
Harvesting and marketing costs				
Picking, packing and hauling		650.00	2250.00	2900.00
Total variable/establishment costs	$/acre	4028.00	5692.00	9720.00
Gross margin		1428.00	3308.00	1880.00
Annual fixed costs	$/acre			
Land (rental price)		500.00	500.00	1000.00
Property tax		50.00	52.00	102.00
Other overhead charges		250.00	250.00	500.00
Total fixed costs	$/acre	800.00	802.00	1602.00
Total costs	$/acre	4828.00	6494.00	11,322.00
Net returns	$/acre	2228.00	2506.00	278.00

Source: Adapted from Evans, E. A., F. H. Ballen and J. H. Crane, 2012, Cost estimates of establishing and producing papaya (*Carica papaya*) in South Florida, Food and Resource Economics Department, Florida Cooperative Extension Service, Institute of Food and Agricultural Sciences, University of Florida, Gainesville, Florida, FE 918:1–5, http://edis.ifas.ufl.edu/.

of Rs. 5,03,500/ha (price of Rs. 5.00/kg). West Bengal farmer (Ranjan Das) obtained a yield of 150 t/ha with a net profit of Rs. 4,20,000/ha. He mainly harvested premature fruits and sold his crop as vegetable at a price of Rs. 3.5/kg. The economics of papaya cultivation has reported a profit of 0.42 million rupees (ha^{-1}) (Biswas 2010). The papaya-based cropping systems (sequential and intercropping) are found most remunerative as they give high net returns (Rs. 1,67,000/ha) in the case of papaya + tobacco intercropping in North Bihar in 1 year 8 months as reported in Table 15.2 (Singh et al. 2010).

TABLE 15.2
Seed Yield and Cost of Production in Gynodioecious and Dioecious cvs., of Papaya with Hand Pollination

Variety	Cross	No. of Buds Crossed/ Plant	No. of Female Parent Plants/ha	No. of Crossed Fruits/ha	No. of Seeds/ Fruit	Weight of Seeds/ Fruit (g)	1000-Seed Weight (g)	Seed Yield/ Plant (g)	Seed Yield (kg/ha)	Total Cost of Seed Production (Rs./ha)	Cost of Seed Production/ kg (Rs.)
Pusa Delicious	F[a] × H[c]	20	1250	25,000	233	4.1	17.6	82.0	102.5	21,875.00	213.40
Pusa Majesty	F × H	20	1250	25,000	194	2.1	10.8	42.0	52.5	21,875.00	416.60
Pusa Giant	F × M[b]	18	2188	39,384	472	7.3	15.5	131.4	295.7	22,363.00	75.60
Pusa Dwarf	F × M	30	2188	65,640	404	5.8	14.3	174.0	391.7	23,938.00	61.10

[a] F, Female.
[b] M, Male.
[c] H, Hermaphrodite.

It had been found that the cultivation of papaya involves high investment, but it is an economically profitable and financially viable fruit crop. Sagar et al. (2012) reported that the average total cost of cultivation of papaya was Rs. 1,76,660 and on an average cost-A (paid out cost) formed 62% of total cost, while cost B accounted for 87% of the total cost. The average farm harvest price received by the papaya growers was Rs. 607 per quintal. On an average, gross income and net returns per hectare were Rs. 4,92,025 and Rs. 3,15,365, respectively. The average per hectare farm business income, family labour income and farm investment income were to the tune of Rs. 3,83,126, Rs. 3,37,653 and Rs. 3,60,838, respectively, on the sample farms. The overall input to output ratio was found as 1:2.79 on the basis of cost incurred (C-2). The average cost of production of papaya was about Rs. 218.00/q which lowers the market price of papaya ranging from Rs. 500 to 700/q. Bulk line cost was Rs. 276/q for papaya at 85% production which covered 69% of farms and 81% of area. On an average, high cost of planting material, lack of knowledge about fertiliser application, the absence of regulated markets, non-availability of high-yielding variety seeds of papaya, lack of knowledge about identifying pests, non-availability of fertiliser in time, long distance of market, non-availability of labour in time were the major production and marketing constraints faced by the papaya growers of middle Gujarat. This criterion indicates the return per rupee invested in papaya enterprise and a wise investor always expects a higher ratio. The benefit to cost ratio in this study was found to be 2.79, which indicates that each rupee invested in papaya enterprise yields Rs. 2.79 as return. Thus, it could be concluded that investment in papaya orchard was economically feasible and financially viable (Devi and Saran 2014).

16 Papaya as Medicaments

There are numerous medicinal uses of papaya for the cure of different diseases. Traditional medicine offers an alternative solution and could be explored as a safer treatment option (Chawla et al. 2014). Various records on the traditional knowledge of papaya for treatment of various diseases by different tribal communities or rural farmers are reported. Papaya has a short growth period and is of great nutritive and health-care values (Li et al. 2012).

16.1 PAPAYA PARTS AND THEIR MEDICINAL USES

Various parts of papaya, namely, ripe and unripe fruits, seeds, bark, leaves, roots and latex are used by the farmers for treatment of several diseases. In folk medicine, latex is used for the treatment of boils, warts, freckles, abortion, dengue, ringworm, expel roundworms, salt making, relieve *asthma*, stomach troubles, purgative, treatment for genito-urinary ailments, tumour destroying, making herbal tea, digestive, aid in chronic indigestion, weight loss, obesity, arteriosclerosis, wound dressing, urinary complaint, anti-haemolytic activity, snake bite to remove poison, high blood pressure, blood purifier and weakening of heart, and so on (Table 16.1). If you are a chain smoker or frequently exposed to second-hand smoke, eating vitamin A-rich foods such as papaya should help in keeping your lung healthy and save life (Aravind et al. 2013).

16.1.1 Fruit

Papaya fruit may be used as a diuretic, anthelmintic, to treat bilious conditions, to combat dyspepsia, other digestive disorders and a liquid portion has been used to reduce enlarged tonsils. In addition, the juice is used for warts, cancers, tumours, corns and skin defects. Danielone is a phytoalexin found in the papaya fruit. This compound showed high antifungal activity against *Colletotrichum gloesporioides*, a pathogenic fungus of papaya. In Asia and Africa, it is applied on the uterus as an irritant to cause abortion. The unripe fruit is sometimes hazardously ingested to achieve abortion. Papaya latex is obtained by cutting the green fruit surface with containers over a couple of days. The latex is then sun dried or oven dried and ground into powder. A proteolytic enzyme, papain, is purified from papaya latex and used in the food and feed industries and also in the pharmaceutical and cosmetic industries (OGTR 2008). Papain is used in food processing to tenderise meat, clarify beer and juice, produce chewing gum, coagulate milk, prepare cereals and produce pet food (Morton 1987). The latex of the papaya plant and its green fruits contains two proteolytic enzymes, papain and chymopapain. The latter is most abundant but papain is twice as important. The lanced fruits may be allowed to ripen and can be eaten locally, or they can be employed for making dried papaya 'leather' or powder

TABLE 16.1
Nutritive Value and Medicinal Uses of Different Parts of Papaya Tree

Source	Nutritional Value	Medicinal Uses	Remarks
Fruits	Ripe fruits Protein, fat, fibre, carbohydrates (glucose, fructose, sucrose), minerals (Ca, P, Fe), Vitamins (ascorbic acid, thiamine, riboflavin, niacin, carotene), amino acids, etc.	Stomachic, digestive, carminative, diuretic, dysentery and chronic diarrhoea, expectorant, sedative and tonic, relieves obesity, bleeding piles, wound of urinary tract, ringworm and skin disease psoriasis.	Excessive use of fruits causes diarrhoea, cold and increases the pain in hydrocele patients (common disease in NEPZ of India).
	Unripe fruits. Citric acid, malic acid, palmitic acid, linolenic acid and glucose.	Laxative, diuretic, dried fruit reduces enlarged spleen and liver, used in case of snakebite to remove poison, abortifaciant, anti- implantation activity and antibacterial activity.	
Seed	Fatty acids, crude protein, crude fiber, oil, carpaine, benzyl isothiocyanate (BITC), benzyl glucosinolate, glucotropacolin, benzylthiourea, hentriacontane, ß-sitostrol, caressing and enzyme myrosin.	Carminative, emmenagogue, vermifuge, abortifacient, counter irritant, in the treatment of ringworm (as paste), psoriasis and anti-fertility agent.	Excessive use of seeds causes bleeding in piles, enlarged liver, anti-fertility, anti-implantation affecting adversely the implantation by BITC.

(*Continued*)

TABLE 16.1 (*Continued*)
Nutritive Value and Medicinal Uses of Different Parts of Papaya Tree

Source	Nutritional Value	Medicinal Uses	Remarks
Root	Carposide and enzyme myrosin.	Abortifacient, diuretic, checks irregular bleeding from uterus and piles, antifungal activity.	Excessive use of root causes bleeding from uterus, piles and antifungal activity.
Leaves	Alkaloids (carpaine, pseudocarpaine and dehydrocarpaine, choline, carposide), vitamin C and E.	Recently matured leaf extract used for improving the platelets count in dengue patients. Fine paste of leaves used for jaundice, urinary complaints, gonorrhoea (infusion), dressing wound due to antibacterial property.	Excessive use of leaf extract causes heart and respiration diseases like digitalis.
Bark	β-sitosterol, glucose, fructose, sucrose and xylitol.	Jaundice, anti-haemolytic and anti-fungal activity.	—
Latex	Proteolytic enzymes, papain, chemopapain, glutamine, cyclortransferase, chymopapains A, B, C, peptidase A, B and lysozymes.	Anathematic, relieves dyspepsia, cures diarrhoea, pain of burn, topical use, bleeding haemorrhoids, stomachic and whooping cough.	Excess use of papain causes allergens, protein breakdown, stomach ulcer, skin burning, etc.

Source: Adapted from Krishna, K. L., M. Paridhavi and J. A. Patel, 2008, *Natural Product Radiance,* 7:364–73; Boshra, V. and A.Y. Tajul, 2013, *Health Environment Journal,* 4(1):68–75; Saran, P. L. and R. Choudhary, 2013, *African Journal of Agriculture Research,* 8(25): 3216–23.

or may be utilised as a source of pectin. Because of its papain content, a piece of green papaya can be rubbed on a portion of tough meat to tenderise it. Sometimes, a chunk of green papaya is cooked with meat for the same purpose. One of the best known uses of papain is in commercial products marketed as meat tenderisers, especially for home use. A modern development is the injection of papain into beef cattle half an hour before slaughtering to tenderise more of the meat than would normally be tendered. Papain-treated meat before eating should be cooked sufficiently to inactivate the enzyme. The tongue, liver and kidneys of injected animals must be consumed quickly after cooking or utilised immediately in food or feed products, as they are highly perishable.

Papain has many other practical applications. People use this in the preparation of different remedies for indigestion. It is used to clarify beer, to treat wool and silk before dying, to de-hair hides before tanning and it serves as an adjunct in rubber manufacturing. It is applied on tuna liver before extraction of the oil, which is thereby made richer in vitamins A and D. It enters into toothpastes, cosmetics and detergents and also pharmaceutical preparations to aid digestion. The papaya eating prevents rheumatism and the latex is used for psoriasis, ringworm and the removal of cancerous growth (Nwofia et al. 2012). Latex (8 g/kg) has been effective in the treatment of ascariosis in pigs in Nigeria; however, mild and temporary adverse effects have occurred in pigs receiving very high doses (Satrija et al. 1994). Papain has been employed to treat ulcers, dissolve membranes in diphtheria and reduce swelling, fever and adhesions after surgery. With considerable risk, it has been applied on meat impacted in the gullet. Chemopapain is sometimes injected in cases of slipped spinal discs or pinched nerves. Precautions should be taken because some individuals are allergic to papain in any form and even to meat tenderised with papain (Saran and Choudhary 2013).

Rubbing the white pulp of raw papaya improves pimples as well as wrinkles. Papaya works as a good bleaching agent. It is an important ingredient in bath soaps, astringents, detergent bars and hand washes. Papayas home recipe can help in removing dead worn-out skin cells and replace it with healthy new cells, thereby lightening the colour of our skin. For this, one can prepare a paste of raw papaya and apply it on the skin once for few days. Ripe papaya fruit is laxative, which assures of regular bowel movement. Ighere Dickson et al. (2012) during the survey reported that the unripe fruit and leaf extract were more frequently used among the people in different places of the Nigeria for medicament of typhoid fever.

16.1.2 Leaf

Papaya leaf has numerous uses. Crushed leaves smeared around tough meat will tenderise it overnight. Leaf juice helps increase white blood cells and platelets, normalises clotting and repairs the liver. The leaf also functions as a vermifuge and a primitive soap substitute in laundering. Dried leaf cured like a cigar is smoked by asthmatic persons for relief. An infusion of fresh papaya leaves is used to expel or destroy intestinal worms. Its infusion is also taken for stomach troubles in Ghana and they say it is purgative and may cause abortion. Packages of dried and pulverised leaves are sold by 'health food' stores for making tea. The leaf decoction is administered as a purgative for horses in the Ivory Coast. It is also used as treatment for genitourinary ailments. The leaf tea or extract has reputation as tumour-destroying agent (Walter 2008). The fresh green tea acts as an antiseptic and dried leaves are best as a tonic and blood purifier (Nwofia et al. 2012). The tea also promotes digestive system and aid in chronic indigestion, weight loss, obesity, arteriosclerosis, high blood pressure and weakening of heart (Mantok 2005). Increasing anecdotal reports of its effects in cancer treatment and prevention, with many successful cases, have warranted that these pharmacological properties be scientifically validated. Seven *in vitro* cell-culture-based studies were reported; these indicate that leaf extracts may alter the growth of several types of cancer cell lines (Nguyen et al. 2013). Leaf extract boosts the production of key signalling molecules called Th1-type cytokines, which help regulate the immune system and inhibition of cancer cell growth. Leaves are used as herbal tea for treatment of malaria. Anti-malarial and anti-plasmodial activities have been noted in some preparations of the plant, but the mechanism is not understood and not scientifically proven (Nakamura et al. 2007). Carpaine, an alkaloid found in papaya leaves, has also been used for medicinal purposes (Sankat and Maharaj 2001). The leaves of papaya plants contain carpain substance, which kills microorganisms that often interfere with the digestive function. Bapedi traditional healers in three districts of Limpopo Province, South Africa, burnt papaya leaves in the consultation hut and patients inhaled the smoke twice to four times a day used for the treatment of tuberculosis (Green et al. 2010). Ethnobotanical studies were conducted to know medicinal plants used by women in the commune of Mahabo-Mananivo of Agnalazaha forest. In some parts of Asia, the young leaves of the papaya are steamed and eaten like spinach. The quantitative phytochemical screening of its leaves aqueous extract revealed the presence of tannins (0.001%), flavonoids (0.013%), saponins (0.022%), phenolics (0.011%), steroids (0.004%) and alkaloids (0.019%), while that of the root gave tannins (0.12%), flavonoids (0.014%), saponins (0.026%), phenolics (0.011%), steroids (0.006%) and alkaloids (0.021%). Cardiac glycosides, anthraquinone, phlobatanin and triterpenes were not detected in the leaves and roots' aqueous extracts of the plant (Bamisaye et al. 2013).

16.1.3 Seed

The black seeds of papaya are edible and have a sharp and spicy taste. Dried papaya seeds actually look quite similar to black pepper and can be used in a similar way. Grinding and spreading a couple over a meal, especially protein-rich meals, is a simple way to add extra enzymes to your diet and improve your digestive health.

Seeds, especially the sarcotesta, are rich source of amino acids. A yellow to brown, faintly scented oil was extracted from the sundried, powdered seeds of unripe papayas at the Central Food Technological Research Institute, Mysore, India. White seeds yielded 16.1% and black seeds 26.8% oil and it was suggested that the oil might have edible and industrial uses. The seeds are used in treatment of sickle cell disease (Imaga et al. 2010). Adding papaya oil and vinegar to bath water, along with essential oils such as lavender, orange and rosemary can be nourishing, refreshing and relaxing, and can work as a pain reliever and muscle relaxant. Air-dried papaya seeds with honey showed a significant effect on human intestinal parasites without significant side effect. Consumption of papaya seed is cheap, natural, harmless, readily available mono-therapeutic and prevents against intestinal parasitosis, especially in tropical communities (Okiniyi et al. 2007). Papaya seed extract may have toxicity-induced kidney failure. Evidently a kidney-transplant patient in London was cured of a post-operative infection by placing strips of papaya on the wound for 48 h. Seeds seem to have more potent medicinal values, namely, antibacterial properties and are effective against *E. coli*, *Salmonella* and *Staphylococcus* infections, protect the kidneys from toxin-induced kidney failure, eliminate intestinal parasites, detoxify the liver, skin irritant to lower fever, cure for piles and typhoid, anti-helminthic and anti-amoebic properties, and so on.

Anti-fertility, anti-implantation and abortifacient properties of extracts from papaya seed have been proved (Chinoy et al. 2006). It has been established in males that the seeds are potential anti-fertility drugs (Lohiya et al. 2005). Seeds are used to produce an Indigenous Nigerian food condiment called '*daddawa*', the Hausa word for a fermented food condiment (Dakare 2004). Fermented seeds have no effects on litters of rats (Abdulazeez et al. 2009), whereas those effects were apparent when the unfermented extract was administered (Abdulazeez 2008). Normal consumption of ripe papaya during pregnancy may not be dangerous; however, unripe or semiripe papaya (which contains high amount of latex that produces marked uterine contraction) could be unsafe for consumption (Krishna et al. 2008).

Antihelmintic activity of papaya seed has been predominantly attributed to carpaine and carpasemine (benzyl thiourea). Carpaine has an intensively bitter taste and a strong depressant action on health. It is present not only in papaya fruit and seed but also in its leaves. Benzyl isothiocyanate (BITC), the main bioactive compound in *C. papaya* seeds (Kermanshai et al. 2001), has been shown to be responsible for the anti-fertility effect (Adebiyi et al. 2003). BITC is capable of damaging the endometrium, making the uterus non-receptive and, thus, affecting adversely the implantation (Adebiyi et al. 2003).

16.1.4 Peel

Papaya peel is often used in cosmetics. It can also be used in many home remedies. The presence of vitamin A helps to restore and rebuild damaged skin. Applied papaya peel is used as a skin-lightening agent. When peel is mixed with honey and applied, it calms and moisturises the skin. The papaya vinegar with lemon juice can be applied to the scalp for 20 min prior to shampooing to fight dandruff.

16.1.5 Roots

The root is ground to a paste with salt, diluted with water and given as an enema to induce abortion. A root decoction is claimed to expel roundworms. Roots are also used to make salt. Aqueous root extract is richer in phytochemical substances as compared to the aqueous leaves extract. Various phytochemical compounds detected are known to have beneficial uses in industries and medical sciences, and also exhibit physiological activity. Therefore, root extract may be a better source for the industrial production and extraction of these phytochemicals, which may serve ethnobotanical uses (Bamisaye et al. 2013). Root is said to help in removing tumours of uterus in animals. Root infusion is also used for syphilis. Juice from papaya roots is used in some countries of Asia to ease urinary troubles. A decoction formed by boiling the outer part of the roots of the papaya tree is used in the cure of dyspepsia.

16.2 SCOPE FOR PHARMACOLOGICAL INDUSTRY

Plant parts including leaf, seed, root and fruit exhibited to have a medicinal value. The stem, leaf and fruit of papaya contain plenty of latex. The latex from unripe papaya fruit contains enzymes papain and chymopapain; other components include a mixture of cysteine endopeptidases, chitinases and an inhibitor of serine protease. Phytochemical analysis of leaf extract revealed the presence of alkaloids, glycosides, flavanoids, saponins, tannins, phenols and steroids. Here the focus is on different properties of papaya such as antioxidant and free radical scavenging activity, anticancer activity, anti-inflammatory activity, treatment for dengue fever, anti-diabetic activity, wound-healing activity and anti-fertility effects. Leaf, fruit and seed are widely used by women for treating headaches, wounds, menstrual pain, stomach ulcer, constipation, indigestion, boil, cysticercosis, toxoplasmosis, tooth decay, cough, improve breastfeeding and yellow fever as a first line of health care for rural families (Razafindraibe et al. 2013). Some of these parts are known to be analgesic, amoebicidic, antibacterial, cardiotonic, cholagogue, digestive, emenagogue, febrifuge, hypotensive, laxative, pectoral, stomachic and vermifugic (Boshra and Tajul 2013). Thus, papaya acts as a multi-faceted plant. It is also imperative to identify the mechanism of the plant compounds and studying the active principle of the extract. Thus, we should include papaya in our diet as fruit salads, fruit juice, leaf extract, decoction prepared through papaya leaves, and so on. The concentrations (of some trace metals such as Zn in seeds and Mn in leaves) were found to be below the maximum acceptable level. In general, the nutritional (major, minor and toxic elements) analysis of papaya seed and leaves has recommended them as good sources of major and minor elements and also free from toxic metals, namely, Cr, Ni, Cd, Co, Pb, and so on (Tigist et al. 2014). This section would focus on potential medicinal properties and utilisation for drug preparations by industries as given below.

16.2.1 Allergens and Side Effects

The flower's pollen has induced severe respiratory reactions in sensitive individuals (Blanco et al. 1998). Papaya pollen in papaya-cultivating areas can contribute to aeropollen and aeroallergen loads (Chakraborty et al. 2005). Papaya contains four cysteine endopeptidases including papain, chymopapain, glycylendopeptidase and caricain. Papain is commonly found in papaya latex (Azarkan et al. 2003). The recorded level of papain in papaya latex is 51,000–135,000 mg/kg (OGTR 2008). Papain can also induce IGE-mediated allergic reactions through oral, respiratory or contact routes of exposure. The typical symptoms include bronchial asthma, rhinitis or both (Van Kampen et al. 2005). One case of a life-threatening anaphylaxis due to occupational exposure to papain was also reported (Freye 1988). Thereafter, such people react to contact with any part of the plant and to eating ripe papaya or any food containing papaya, or meat tenderised with papain. People who eat too much papaya and ingest high levels of papain may develop symptoms consistent with hay fever or asthma, including wheezing, breathing difficulties and nasal congestion.

Papaya releases a latex fluid when not quite ripe, which can cause irritation and provoke allergic reaction in some people. Externally the papaya latex is an irritant to the skin and internally it causes severe gastritis due to high fibre content of papaya and also contributes to unrest of the digestive system. The latex of the fruit's skin can also cause irritation of the stomach. The latex concentration of unripe papayas is speculated to cause uterine contractions, which may lead to a miscarriage. Skin irritation is caused to papaya harvesters because of the action of fresh papaya latex and possible hazard of consuming undercooked meat tenderised with papain. Papaya is frequently used as a hair conditioner, but should be used in small amounts (Saran and Choudhary 2013). Some people are allergic to various parts of the fruit and even the enzyme papain has its negative properties. Papaya seed extracts in large doses have a contraceptive effect on rats and monkeys, but in small doses have no effect on the unborn animals. Excessive consumption can cause carotenemia, the yellowing of soles and palms, which is otherwise harmless. Papaya contains about 6% of the level of beta carotene found in carrots, which is the most common cause of carotenemia (Bamisaye et al. 2013).

16.2.2 Antioxidants and Free Radical Scavenging Activity

Phytochemical analysis of leaf extracts revealed the presence of alkaloids, glycosides, flavanoids, saponins, tannins, phenols and steroids (Gill 1992; Owoyele et al. 2008). The stems, leaves and fruits of papaya contain plenty of latex. The latex of papaya is a rich source of four cysteine endopeptidases namely papain, chymopapain, glycylendopeptidase and caricain—a papaya endopeptidase II (Azarkan et al. 2003). As the fruit ripens, papain and chymopapain get degraded and are not present in the ripe fruit (Oloyede 2005). Other components include a mixture of cysteine endopeptidases such as endopeptidase IV, omega endopeptidase, class-II and class-III chitinase and an inhibitor of serine protease (Odani et al. 1996; Azarkan et al. 1997; El Moussaoui et al. 2001). The aqueous extract of unripe *C. papaya* administered orally in Wistar albino rats demonstrated no adverse effect on the histology of liver, kidney, heart and small intestine (Oduola et al. 2010).

The leaves, seeds and juice show free radical scavenging and antioxidants activity. The antioxidant activity of various fractions (ethanol, petroleum ether, ethyl acetate, *n*-butanol and aqueous extract) from seeds was evaluated and showed that ethyl acetate and *n*-butanol fractions demonstrated antioxidant and free radical scavenging activity than other fractions (Zhou et al. 2011). Papaya juice is an efficient scavenger of highly reactive hydroxyl radicals (Webman et al. 1989), which significantly decreased the lipid peroxidation levels and increased the antioxidant activity in rats (Mehdipour et al. 2006). The leaf extract of *C. papaya* evidenced significant antioxidant and free radical scavenging potential (Okoko and Ere 2012). The peroxidase is present in the unripe fruit but it is gradually decreased after fruit ripening (Pandey et al. 2012). The pulp of papaya is rich in benzyl glucosinolate in the premature stage, which is present in the seed after fruit ripening (Li et al. 2012). The benzyl glucosinolate is hydrolysed to benzyl isothiocyante (BITC). The seed extract demonstrated a rich source of BITC (Nakamura et al. 2007). It shows that papaya is an excellent source of vitamin A, vitamin C and dietary fibre and a good source of vitamin E.

16.2.3 Anti-Diabetic Activity

The aqueous extract of leaves significantly reduced plasma blood glucose level and serum lipid profile in diabetic rats (Juárez-Rojop et al. 2012; Maniyar and Bhixavatimath 2012). The ethanolic extract of leaves demonstrated significant reduction in the blood glucose level and regeneration of the beta cells of pancreas in diabetic mice (Azarkan et al. 2003). The aqueous extract of unripe fruit significantly inhibited the key enzymes α-amylase and α-glucosidase involved in diabetes and also inhibited the lipid peroxidation in rat pancreatic cells studied *in vitro* (Oboh et al. 2013).

16.2.4 Anti-Infertility Activity

The seeds were shown to have antifertility properties in male albino rats. Seed extract treated in male albino rats reduced the cauda epidymal and testicular sperm counts (Lohiya and Goyal 1992). Male wistar rats treated orally with seed extract (200 mg/kg) demonstrated hypertrophy of pituitary gonadotrophs and gradual degeneration of germ cells, sertoli cells and leydig cells of testis thereby, drastically affecting the male reproductive functions (Udoh et al. 2005). The aqueous extract of papaya seed administered to male Sprague–Dawley rats suppressed the steroidogenic enzymes in the testis and reversible changes occurred when the extract was withdrawn after 30–45 days of treatment (Uche-Nwachi et al. 2011). The papaya seed extract can also be used as an effective male contraceptive (Chinoy et al. 1994). Women in India, Bangladesh, Pakistan, Sri Lanka and other countries have long used green papaya as a herbal medicine for contraception and abortion. Enslaved women in the West Indies were noted for consuming papaya to prevent pregnancies and, thus, preventing their children from being born into slavery.

16.2.5 Anti-Inflammatory Activity

The papaya leaf extract was examined in rats using oedema, granuloma and arthritis models. The extract showed significant reduction in paw oedema, granuloma formation and reduced inflammation in rats (Owoyele et al. 2008). Intake of papaya fruits in healthy individuals alleviated anti-inflammatory response mediated through regulatory T-cells (Abdullah et al. 2011). However, the latex obtained from unripe fruit has the property of inducing inflammation and it has been proved in rat as a model for testing the anti-inflammatory activity of compounds (Gupta et al. 1992). Protein enzymes including papain and chymopapain and antioxidant nutrients found in papaya, including vitamin C, vitamin E and beta-carotene, reduce the severity of the conditions such as asthma, osteoarthritis and rheumatoid arthritis.

16.2.6 Antibiotic Activity

The extracts of ripe and unripe papaya fruits and of the seeds are active against Gram-positive bacteria. Strong doses are effective against Gram-negative bacteria. The substance has protein-like properties. The fresh crushed seeds yield the aglycone of glucotropaeolin benzyl isothiocyanate (BITC), which is bacteriostatic, bactericidal and fungicidal. A single effective dose is 5 g seeds (25–30 mg BITC) for the same. Papaya was also found to be effective in curing postoperative infection in a kidney transplant patient by strip laid on the wound and left for 48 h.

16.2.7 Anti-Cancer Activity

The fibre of papaya is able to bind cancer-causing toxins in the colon and keep them away from the healthy colon cells. These nutrients provide synergistic protection for colon cells from free radical damage to their DNA (Desser et al. 2011). Men consuming fruits were 82% less likely to have prostate cancer compared to those consuming the least lycopene-rich foods. An *in vitro* study using the extract of the seed homogenate was highly effective in inhibition of superoxide generation and inducing apoptosis in acute promyelocytic leukaemia cell line HL-60 and the activity was mainly contributed by benzyl isothiocyanate (BITC) (Nakamura et al. 2007). The BITC isolated from the papaya fruit extract induced cytotoxic effect in proliferating human colon CCD-18Co cells to the quiescent state (Miyoshi et al. 2007). The aqueous extract of *Carica papaya* exposed to onion bulbs disturbed the mitotic cell division of *Allium cepa* by affecting the spindle formation and hence shows its cytotoxic effect (Akinboro and Bakare 2007). The aqueous extract of papaya flesh (0.01%–4.00% v/v) treated with breast cancer cell line, MCF7 revealed significant inhibition of cell proliferation (García-Solis et al. 2009). The aqueous extract of papaya leaves in an unrevealed composition is shown to possess anticancer activity and inhibition of cell proliferation in a variety of cancer cell lines, which has been patented by Morimoto et al. (2008). Likewise, the aqueous extract demonstrated antitumour activity and immunomodulatory activity in tumour cell lines and it proved up regulation of immunomodulatory genes by microarray studies (Otsuki et al. 2010). However, further investigation using cell culture studies, animal studies and clinical trials is needed for proving the chemoprevention and therapeutic potential of different papaya plant parts and check the adverse effects, if any, in consumption of some parts of papaya (Nguyen et al. 2013).

16.2.8 Anti–Rheumatoid Arthritis Activity

Vitamin C-rich foods, such as papaya, provide humans with protection against inflammatory polyarthritis, a form of rheumatoid arthritis involving two or more joints (Aravind et al. 2013).

16.2.9 Anti-Dengue Activity

A case report from Pakistan (Ahmad et al. 2011) documented that aqueous extract of *C. papaya* leaves administered to a patient affected with dengue fever twice daily for five consecutive days exhibited elevated platelets count from $55 \times 10^3/\mu L$ to $168 \times 10^3/\mu L$. Another study in the murine model also evidenced an increase in platelets and RBC count without any acute toxicity after oral administration of papaya extract (Dharmarathna et al. 2013). The juice prepared from leaves recorded a significant increase of platelet count in a randomised controlled trial conducted on patients with dengue fever and dengue haemorrhagic fever (Subenthiran et al. 2013).

16.2.10 Anti-Heart Attack

The folic acid found in papayas is needed for the conversion of homocysteine into amino acids such as cysteine or methionine. If unconverted, homocysteine can directly damage blood vessel walls and is considered a significant risk factor for a heart attack or stroke (Aravind et al. 2013).

16.2.11 Anti-Nutrients and Toxicants Activity

Peel and pulp of ripe papaya fruits contain low amounts of antinutritional factors such as tannin (10.16 mg/100 g of dry matter), phytate (3.29 mg/100 g of dry matter) and oxalate (1.89 mg/100 g of dry matter) creating incompatibility problems as reported by Onibon et al. (2007). Carpaine is a major alkaloid found in various parts of papaya, but is primarily found in leaves (Morton 1987; Duke 1992; Krishna et al. 2008). The major natural toxicants found in papaya are benzyl glucosinolate (BG), benzyl isothiocyanate (BITC) and alkaloids. Fruit and seed extracts have pronounced bactericidal activity. The seeds of unripe fruits are rich in benzyl isothiocyanate, a sulphur containing chemical that has been reported to be an effective germicide and insecticide. These substances are important for plant natural defence mechanisms (El Moussaoui et al. 2001). Although both BG and BITC are found in papaya peel, pulp and seed, the highest levels of BG and BITC are found in seeds, 1269.3 and 461.4 μmol/100 g fresh weights, respectively. The levels of BG and BITC in papaya pulp were <3.0 μmol/100 g fresh weight (Nakamura et al. 2007). The concentration of BITC decreases in pulp and increases in seeds during fruit ripening.

16.2.12 Wound-Healing Activity

The aqueous extract of *C. papaya* significantly enhances the wound healing that makes it an ideal dressing component for treatment of wounds (Mahmood et al. 2005). Fruits and seeds of *C. papaya* were evaluated for wound healing activity using a wound excision model in diabetic rats, which showed significant reduction in the wound area compared to untreated diabetic control. It also showed increased granulation, elevated hydroxyproline content and deposition of collagen in the wound area (Nayak et al. 2007, 2012). Papaya latex prepared in carbapol gel for treatment of burns demonstrated a significant increase in hydroxyproline content as well as wound contraction in Swiss albino mice (Gurung and Skalko-Basnet 2009). Diabetic mice supplemented with fermented papaya preparation (FPP) showed effective recruitment of monocytes and proangiogenic response by the macrophages at the wound site resulting in wound closure (Collard and Roy 2010).

16.2.13 Skin Discolouration Activity

Eating too much of a yellow, green or orange-coloured food that contains beta carotene can cause a benign form of skin discolouration called carotenemia. The palms of the hands and soles of the feet are the most visible areas of the body affected by carotenemia. Cutting back on your papaya consumption will resolve the discolouration of the skin (Aravind et al. 2013).

16.2.14 Anti-Coagulant Activity

Injection of papian extract in a dog increases prothrombin and coagulation threefold. It is also claimed that the enzyme eliminates necrotic tissues in chronic wounds, burns and ulcers. Papain is also of commercial importance in the brewery, food and textile industries (Aravind et al. 2013).

16.2.15 Folk Uses for Major Disease Management

Several farmers, tribes and other communities of Africa, Nigeria, Ghana, India, Australia, Pakistan, Sri Lanka, and so on, are using the papaya parts for the treatment of different diseases. In rural areas of Afro-Asian countries, medical specialists are not easily available. Thus, patients try to manage different diseases by local *Ayurvedacharya*.

Diagnostic surveys were conducted using both interaction and participatory rural appraisal (PRA) techniques at Indian Agricultural Research Institute Regional Station, Pusa, Samastipur, Bihar (India). The investigation included individual and group interviews with 800 respondent farmers with a mission to identify the indigenous treatment practices. After locating the indigenous practices, a check list of 12 diseases/disorders was prepared (Table 16.2). Among different diseases, farmers put jaundice (94.96%) and stomach problems (87.66%) at rank I and dengue (67.38%) at rank II. Farmers between 46 and 60 years age group possess more traditional knowledge (91.70%) about the medicament properties of papaya parts and their role in treatment of different common diseases (Saran et al. 2015).

Traditional knowledge regarding medicinal uses of different parts of papaya and management of several common diseases by their use depends upon the age group of farmers and tribals. The knowledge level of farmers/tribals increases with the increase in age. Experienced people have more traditional knowledge about the medicament properties of papaya parts and their role in managing different common diseases (Figure 16.1). The farmers in the age group, 46–60 years, possess more knowledge (91.70%) about the medicinal properties of different parts of papaya, followed by the age group, 31–45 years (33.3%) and ≤30 years age group (16.70%) (Saran et al. 2014). Similarly, the age group between 45 and 86 years is more proactive in use of medicinal plants in the management of diabetes mellitus (Vidhyasagar and Murthy 2013). People aged 40–49 years have the highest frequency of use of medicinal plants (98.29%). This age group was followed by the 50–59-year-old age bracket (96.15%), the 30–39-year-old age bracket (94.59%), 60 years and older bracket (89.36%), the 20–29-year-old bracket (86.91%) and finally the youngest bracket,

TABLE 16.2
Relative Importance of Various Disease Management Categories Perceived by Respondents

	Frequency of Adopter Farmers	
Disease	**Score (%)**	**Rank**
Abortifaciant, anti-implantation (abortion)	00.78[h]	IV
Anti-haemolytic activity	22.51[f]	IV
Dengue fever	67.38[c]	II
High blood pressure (heart patients)	02.21[h]	IV
Jaundice	94.96[a]	I
Ringworm	26.30[e]	III
Roundworm	27.00[e]	III
Snakebite to remove poison	01.22[h]	IV
Stomach problems (digestive, carminative, dysentery and chronic diarrhoea)	87.66[b]	II
Urinary complaints (diuretic)	44.15[d]	III
Weight loss	09.07[g]	IV
Wound dressing	25.83[e]	III

Source: Adapted from Saran, P. L. et al., 2015, *Indian Journal of Traditional Knowledge,* 14(1): (in press).

Note: Means with the same letter (superscript) in the columns show no significant difference ($P = 0.05$) based on the Duncan Multiple Range Test.

15–19 years old at 45.73%. People at least 30 years old have increased knowledge in terms of medicinal plants, while lower knowledge levels occur in the younger age groups (Razafindraibe et al. 2013).

16.3 THERAPY CASES

Papaya has several antibiotic, allergic, anti-nutritional and toxic properties. Untested herbal medicines could be potentially injurious to human health (Saran and Choudhary 2013). Many plants used in traditional and folk medicines are potentially toxic, mutagenic and carcinogenic (Dharmarathna et al. 2013). Many biologically active phytochemicals from different parts of papaya tree (latex, seed, leaf, root, stem, bark and fruit) have been isolated and studied for their potency.

16.3.1 Case Studies

Papaya fruit is a rich source of nutrients such as provitamin-A (carotenoids), vitamin B, vitamin C, lycopene, dietary minerals and dietary fibre. Biochemical

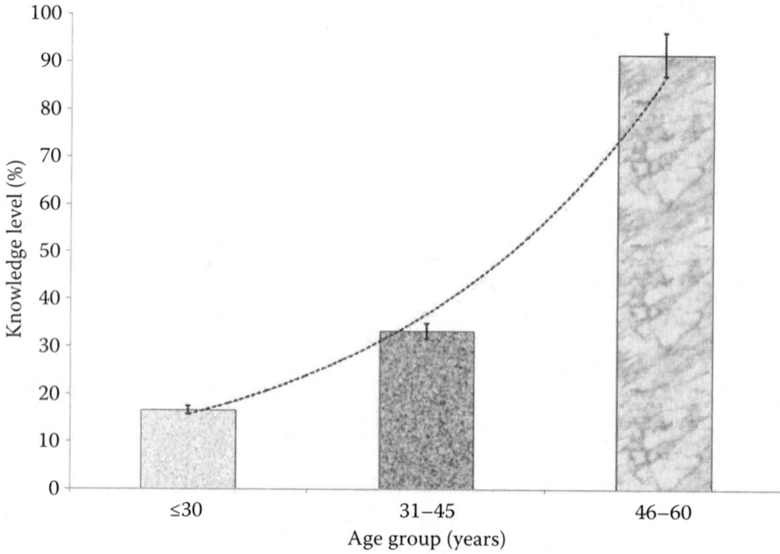

FIGURE 16.1 Relationship between the knowledge level regarding medicinal uses and age of farmers. (Adapted from Saran, P. L. et al., 2015, *Indian Journal of Traditional Knowledge* (in press).)

effects of unripe fruit pulp on alloxan-induced diabetes in rats have been reported. Animals weighing (160–200 g) were divided into three groups of 10 animals each: Group I (normal control), Group II (diabetic control) and Group III (test control). Diabetes was induced in albino rats by intra-peritoneal injection of alloxan monohydrate at a single dose of 120 mg/kg body weight into groups II and III, respectively. Animals in groups I and II received normal feeds, while animals in group III were fed with unripe pulp for a period of 28 days. Body weight and glucose levels were measured on days, 0, 7, 14, 21 and 28. It was observed in this study that there was a significant reduction of glycated haemoglobin in the test group when compared with the induced control (diabetic rats) and normal control group (Figure 16.2).

The unripe pulp elicited significant reduction of blood glucose (Figure 16.3), lipid profile parameters, except high-density lipoprotein cholesterol (HDL-C), which significantly increased. Results also showed significant reduction in body weight (Figure 16.4). This action is believed to be due to the bioactive constituents of the plant (Ahamefula Sunday et al. 2014). Unripe pulp exhibits proven potentials in the parameters evaluated to become of important medicinal and pharmacological interest.

The dengue fever attack cases are reported from the onset of rainy season to early winter season in India, especially North Eastern Plains Zone. The dengue fever recurs every year and causes several deaths in this region. Dengue is a common pathogenic disease often proving fatal, more commonly affecting the people in tropical and subtropical areas. Aedes mosquito is the vector for this disease, and outbreaks of dengue

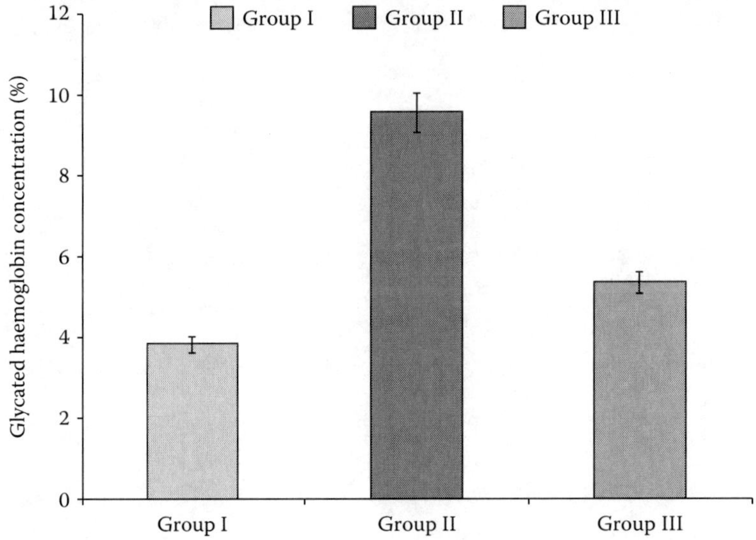

FIGURE 16.2 Effects of unripe pulp of *Carica papaya* on glycated haemoglobin levels. (Adapted from Ahamefula Sunday, E., E. Ify and K. O. Uzoma, 2014, *Journal of Pharmacognosy and Phytochemistry,* 2(6): 109–14.)

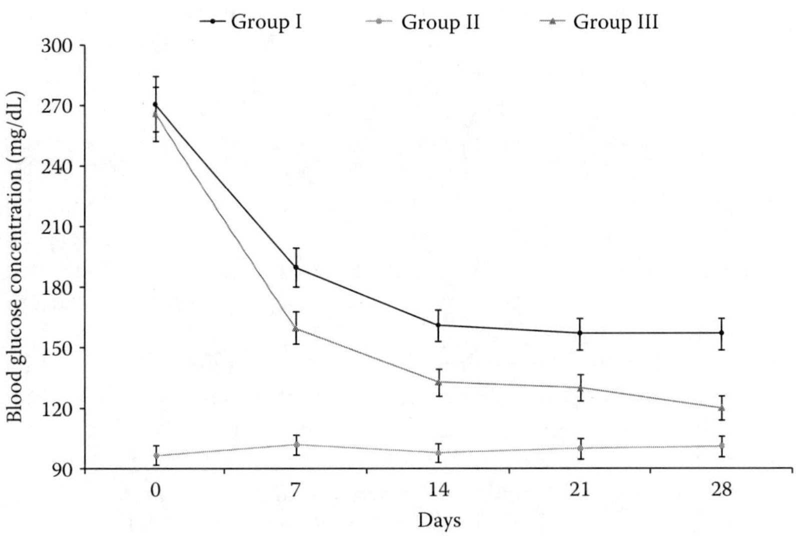

FIGURE 16.3 Showing the effects of unripe pulp of *Carica papaya* on blood glucose level. (Adapted from Ahamefula Sunday, E., E. Ify and K. O. Uzoma, 2014, *Journal of Pharmacognosy and Phytochemistry,* 2(6): 109–14.)

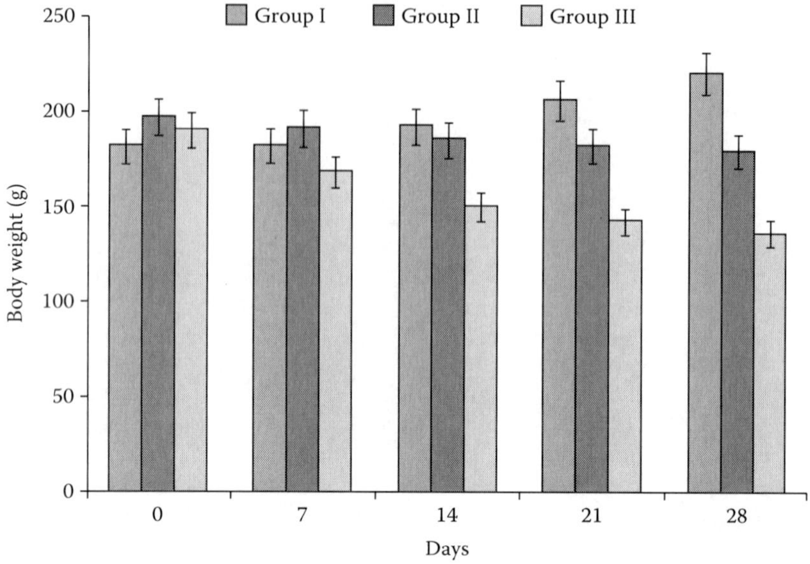

FIGURE 16.4 Effects of unripe pulp of *Carica papaya* on body weight. (Adapted from Ahamefula Sunday, E., E. Ify and K. O. Uzoma, 2014, *Journal of Pharmacognosy and Phytochemistry*, 2(6): 109–14.)

often cause endemic damage to life. Supplement of papaya unripe fruit and leaf extract on platelets count for the dengue patient as supportive therapy was given with medical treatment. Traditional use of papaya leaf in the treatment of dengue fever (DF) has also been reported (Saran and Choudhary 2013). Dengue patients were administered 50 mL of papaya fruit extract and 20–25 mL of papaya leaf extract in addition to medical management. Papaya fruit and leaf extract were administered before food, once a day for four consecutive days with sugar to avoid bitterness and for better palatability (Table 16.3). Fresh papaya mature leaves and unripe fruits (90–100 days old) were grinded with mortar pestle and juice mixer, respectively, without adding water. The pastes were kept in muslin cloth. After squeezing, the filtrate was collected and stored in a container. All aseptic measures were taken care of while preparing the extract.

Similarly, five days of oral administration of 25 mL aqueous extract of papaya leaves to the patients twice daily has been reported (Ahmad et al. 2011). The aqueous extract of leaves of this plant exhibited potential activity against dengue fever by increasing the platelet (PLT) count (Table 16.4) (Ahmad et al. 2011; Dharmarathna et al. 2013). The bioactive anti-malarial element has important commercial potential since the fruit grows in abundance in the tropics (Krishna et al. 2008).

Jaundice and hydrocele are significant causes of morbidity and mortality and lead to adverse effects on human inhabitants in several regions of India, especially in NEPZ. Many biologically active phytochemicals from different parts of papaya tree, namely, leaf, stem bark and flower have been isolated and studied for their potency

TABLE 16.3
Response of Papaya Unripe Fruit Extract and Leaf Extract as Supportive Therapy (ST) on Platelet Count in Dengue Patients

	Platelet Count				
Treatment	**Initial**	**1st day**	**2nd day**	**3rd day**	**4th day**
Medical treatment	$61 \times 10^3/\mu L$	$70 \times 10^3/\mu l^b$	$97 \times 10^3/\mu L^a$	$122 \times 10^3/\mu L^b$	$170 \times 10^3/\mu L^c$
Fruit extract (ST)	$62 \times 10^3/\mu L$	$74 \times 10^3/\mu L^a$	$110 \times 10^3/\mu L^a$	$130 \times 10^3/\mu L^b$	$186 \times 10^3/\mu L^b$
Leaf extract (ST)	$58 \times 10^3/\mu L$	$77 \times 10^3/\mu L^a$	$100 \times 10^3/\mu L^a$	$154 \times 10^3/\mu L^a$	$202 \times 10^3/\mu L^a$

Source: Adapted from Saran, P. L. et al., 2015, *Indian Journal of Traditional Knowledge*, 14(1): (in press).

Note: Means with the same letter (superscript) in the columns show platelet counts not significantly different (P = 0.05) based on the Duncan Multiple Range Test.

against jaundice (Table 16.5). Papaya leaves have also been used as treatment for genito-urinary ailments (Saran and Choudhary 2013). The unripe fruit is also used traditionally for treating jaundice by the Yoruba tribe of Nigeria (Elujoba 2001) and farmers in India (Devi and Saran 2014).

The labourers, especially tribals, are more susceptible to hydrocele as compared to farmers due to lack of education and hygienic conditions (Table 16.6). They have traditional knowledge that eating papaya fruits increase the pain in hydrocele patients. Peel and pulp of ripe papaya fruits contain low amounts of anti-nutritional factors such as tannin, phytate and oxalate creating incompatibility problems. The ripe fruits are also a poor source of benzyl glucosinolate (BG) and benzyl isothiocyanate (BITC) toxicants (Li et al. 2012).

16.4 IS EATING PAPAYA DURING PREGNANCY SAFE?

In India, there are many old wives' tales and information flooding the media whether ladies should or should not eat certain foods during pregnancy. Fruit is widely classified as harmful in pregnancy; hence pregnant women are strictly forbidden from eating it for fear of its teratogenic and abortifacient effects (Adebiyi et al. 2003). There have been many research projects into the effects of foods on pregnant women and papaya is no exception. The problem with papaya is that in an unripe state it contains high concentrations of latex, which gets reduced upon ripening and once the fruit is completely ripe, it has almost no latex left. The papaya latex's main constituents are papain and chymopapain, which have teratogenic (abnormalities of physiological development) and abortifacient effects. Chinoy et al. (2006) proved the anti-fertility, anti-implantation and abortifacient properties of extracts from papaya seeds, which act by increasing the chances of uterine contractions as the papain acts like prostaglandin and oxytocine, which are known to put a mother's body into labour and hence an adverse effect on babies' and mothers' health.

TABLE 16.4
Haematological and Biochemical Parameters in the Groups after Giving Leaf Extract

	Day 1		Day 7		Day 14		Day 21	
Parameter	**CG[a]**	**TG[b]**	**CG[a]**	**TG[b]**	**CG[a]**	**TG[b]**	**CG[a]**	**TG[b]**
Platelet count ($\times 10^5/\mu L$)	3.67 ± 0.16	3.36 ± 0.16	4.52 ± 0.15	9.00 ± 0.35	5.21 ± 0.13	10.86 ± 0.38	5.53 ± 0.12	11.33 ± 0.35
RBC ($\times 10^6/\mu L$)	6.23 ± 0.17	5.87 ± 0.19	5.95 ± 0.18	6.63 ± 0.32	6.61 ± 0.28	7.95 ± 0.59	6.00 ± 0.31	7.97 ± 0.61
WBC ($\times 10^3/\mu L$)	7.45 ± 0.23	7.61 ± 0.13	7.16 ± 0.21	7.62 ± 0.32	7.34 ± 0.15	7.71 ± 0.61	7.52 ± 0.11	8.01 ± 0.42
SGOT (U/L)			88.67 ± 7.60	118.67 ± 25.91	96.17 ± 40.00	110.17 ± 23.00	90.00 ± 16.47	90.00 ± 13.40
SGPT (U/L)			28.50 ± 2.70	24.17 ± 3.70	17.83 ± 4.90	27.67 ± 9.97	47.50 ± 7.40	42.83 ± 3.32
Serum creatinine (mg/dL)			0.12 ± 0.12	0.12 ± 0.02	0.03 ± 0.04	0.1 ± 0.62E-18	0.1 ± 6.2E-18	0.10 ± 0.12
PCV (%)			41.8 ± 0.48	40.83 ± 1.85	43.7 ± 2.72	40.83 ± 1.19	41.00 ± 4.32	44.83 ± 1.79

Source: Adapted from Dharmarathna, S. L. C. A. et al., 2013, *Asian Pacific Journal of Tropical Biomedicines,* 3(9):720–24.
CG[a]: Control group; TG[b]: Test group

TABLE 16.5
Response of Different Supportive Therapies (ST) in Jaundice Patients

Treatment	Respondents	Recovery Duration (days)	Rank
Medical treatment (MT)	120	30[a]	V
MT + Unripe fruit (ST)	72	24[b]	I
MT + Raw fruit latex in *batasa* (ST)	11	25[b]	II
MT + Curd (ST)	19	28[a]	III
MT + Sweet (Rasgulla) (ST)	18	28[a]	IV

Source: Adapted from Saran, P. L. et al., 2015, *Indian Journal of Traditional Knowledge,* 14(1): (in press).

Note: Means with the same letter (superscript) in the columns show recovery durations not significantly different (P = 0.05) based on the Duncan Multiple Range Test.

TABLE 16.6
Response of Eating Raw Papaya Fruit as Promotive Therapy (PT) in Hydrocele Patients

Treatment	Number of Respondents	Hydrocele Patients (%)	Preference for Fruit Eating (%)
Labourers	72	43.06[a]	59.72[b]
Farmers	311	32.80[b]	70.74[a]
Average	—	34.73	68.67

Source: Adapted from Saran, P. L. et al., 2015, *Indian Journal of Traditional Knowledge,* 14(1): (in press).

Note: Means with the same letter (superscript) in the columns showing hydrocele patient and preference for fruit eating (%) do not differ significantly (P = 0.05) based on the Duncan Multiple Range Test.

The latex can also cause marked oedema and haemorrhagic placentas which bleed and haemorrhage from the edge of the placenta and can result in severe complications in pregnancy and result in an early delivery (Aravind et al. 2013). Therefore, ripe fruits may be eaten by pregnant women after removing seed and other placental parts. Pulp of ripe fruit contains low amounts of anti-nutritional factors such as tannin, phytate and oxalate-creating fewer incompatibility problems. The ripe fruits are also poor source of benzyl glucosinolate (BG) and benzyl isothiocyanate (BITC) toxicants (Li et al. 2012).

16.5 POTENTIAL BIOACTIVE COMPOUNDS

Glucosinolates are sulphur-containing secondary metabolites found largely in Brassicaceae family and are derived from amino acid precursors (Fahey et al. 2001). Glucosinolates undergo hydrolysis readily upon cell rupture by cutting, chewing, cooking, fermenting or freezing with the presence of naturally occurring enzyme myrosinase (3-thioglucosidase glucohydrolase) to produce isothiocyanates and simple nitriles (mostly inactive), namely, nitriles, epithionitriles, oxazolidine-thiones and thiocyanates depending upon the structure of glucosinolates and reaction conditions (Blazevic et al. 2010). However, formation of benzyl isothiocyanate depends upon reaction conditions where benzyl nitrile is formed at the expense of benzyl isothiocyanate. Thus, the potential health benefits of benzyl isothiocyanate may be surpassed by the ineffective benzyl nitrile. Isothiocyanates are well known for its diverse biological activities ranging from bactericidal, nematocidal, fungicidal, insecticidal, antioxidant, antimutagenic, antipoliferative and allelophatic properties (Kim et al. 2010). Isothiocyanates are potential anticarcinogen, which could inhibit the liver, lungs, colon, breast, ovary, prostate, bladder and pancreas cancer (Vig et al. 2009). It also induces apoptosis in various cancer cell lines (Kuang and Chen 2004). One of the convincing glucosinolates hydrolysis products in cancer studies is benzyl isothiocyanate, a hydrolysis product of benzyl glucosinolate. Benzyl isothiocyanate is known to induce cell apoptosis in human breast cells (Xiao et al. 2008), pancreatic cancer cells (Wicker et al. 2010; Sahu et al. 2009) and reduced the growth of solid tumours (Kim et al. 2010).

The concentration of benzyl isothiocyanate was reported 52.2 mg/kg, 18.0 mg/kg and 3.6 mg/kg in leaf, unripe fruit and flower, respectively. The highest amount of BITC was reported at room temperature (25°C) and it decreases gradually as the temperature rises up to 80°C. Comparing three common domestic methods of cooking vegetable, that is, blanching, boiling and slow heating, the results show that high temperature treatment produced mainly benzyl nitrile, while slow heating up to 40°C produced more benzyl isothiocyanate (Volden et al. 2009). In these cooking experiments, both hydrolysis products were found largely leached into the cooking soup. Isothiocyanates itself are temperature labile and volatile, which makes it to vanish from heat processed foods (Volden et al. 2008; Volden et al. 2009).

The optimum pH for BITC was reported in the range of 6.0–7.0. Acidic conditions favour nitrile formation and higher pH favour isothiocyanate (Nagappan and Surugau 2011). Small amount of ascorbic acid enhanced the formation of BITC; however, higher concentration inhibits the production. Ascorbic acid (0.3 mM) has enhanced the nitrile production by 11-fold as compared without addition of ascorbic acid (Nagappan and Surugau 2011).

17 Protected Cultivation

Protected cultivation is defined as a cropping technique wherein the micro-environment surrounding the plant body is controlled partially or fully as per the plant's need during its period of growth to maximise the yield and resource saving. With the advancement in agriculture, various types of protected cultivation practices suitable for a specific type of agro-climatic zone have emerged. Among these protective cultivation practices, greenhouse or polyhouse is useful for the hill zones and net house for hot areas for papaya cultivation. The greenhouse is generally covered by transparent or translucent material such as glass or plastic. The greenhouse covered with simple plastic sheet is termed as polyhouse. The greenhouse generally reflects back about 43% of the net solar radiation incident upon it allowing the transmittance of the 'photosynthetically active solar radiation' in the range of 400–700 nm wavelength. The sunlight admitted to the protected environment is absorbed by the crops, floor and other objects. These objects in turn emit long wave thermal radiation in the infrared region for which the glazing material has lower transparency. As a result, the solar energy remains trapped in the protected environment, thus, raising its temperature. This phenomenon is called the 'greenhouse effect'. This condition of natural rise of air temperature in greenhouse is utilised in the cold regions to grow crops successfully. However, during the summer season due to the aforementioned phenomenon ventilation and cooling is required to maintain the temperature inside the structure well below 35°C. The ventilation system can be natural or a forced one. In the forced system fans are used which draw out 7–9 m^3 of air/sec/unit of power consumed and able to provide two air changes/minute.

The use of protected cultivation for fruit crops, particularly papaya, has not been adopted on the same scale as in vegetable- and flower-production systems. The challenges in producing this crop in protected systems are needed particularly where pest and virus are the serious threats to quality and productivity. The objectives of this chapter are to explore the concepts of protected agriculture as they apply to papaya and illustrate the prospects for protected agriculture to address biotic and abiotic challenges to fruit production.

17.1 SCENARIO

Besides sporadic efforts of Defence Research and Development Organization (DRDO) in Ladakh, protected cultivation technology in India for commercial production is hardly three decades old, whereas in developed countries namely, Japan, Holland, Russia, the United Kingdom, China and others, it is about two centuries old. In recent years, Israel is one country which has taken big advantage of this technology by producing quality vegetables, flowers, fruits, etc. in water deficit desert area for meeting not only its small domestic demand but also the huge export demands. Recently, India has established collaboration with Israel for demonstration of this technology at the Center for Protected Cultivation Technology (CPCT), Indian Agricultural Research Institute, New Delhi, which has paid good dividends.

17.2 SCOPE

The technology of protected cultivation mainly of papaya is most suitable to promote much needed peri-urban and urban agriculture being useful to both small-scale and large-scale farmers. The bulging metros would need this technology if they want to have fresh fruits, especially papaya, as transportation of these from distant places, mainly rural areas, would not render them as fresh produce. Moreover, under net house pure seed production and round the year disease free planting material production is carried out successfully. Post-harvest losses can also be reduced significantly when production is located close to consumers, especially of perishables.

It is an upcoming and alternative production system involving high-tech and intensive practices, especially for urban and export demands of crops for fruit, nutrition and economic security. Burgeoning population, fragmentation of land holdings, depletion and erosion of natural resources are adversely affecting agricultural productivity. The protected cultivation offers several advantages to grow high-value crops such as papaya, strawberry, etc. with improved quality even under unfavourable and marginal environments. It has the potential of fulfilling the requirements of small growers as it can increase the yield manifold per unit area. The planting material can also be grown round the year, including off-season with increased profitability. The market size of fruits has increased several folds in the last decade due to globalisation of trade and liberalisation of Indian economy. Besides the market size, the purchasing power of people and consumer profile has also changed positively towards the nutrient-rich quality fruits and major herbs for medicament. The consumption of the fruits is increasing in middle and upper class of the Indian society and consumption of unripe fruit and other parts of plant by lower society as medicament is also increasing. The rising demands from such sectors of the consumers could not be met from the traditional system of fruit production alone. Plants in open field conditions experience short growing season, unfavourable climatic conditions (too cold, too hot, too dry and cloudy) impairing photosynthetic activities and vulnerable to predators, pests, weeds, depleted soil moisture and plant nutrients leading to drastic reduction in fruit yields. Hence, there is a need to protect this valuable crop to sustain the productivity. A breakthrough in production technology that integrates market-driven quality parameters with the production system, besides ensuring a vertical growth in productivity is required. In fruit production, protective structures have proved to be beneficial in crop improvement, plant propagation, protecting from biotic and abiotic stresses, altering cropping season, growth and yield enhancement and quality improvement. It has been used to modify the microclimate to advance maturity, increase yields and expand the area of production.

Protected cultivation is economically more rewarding in production of high value, low volume crops, seeds, planting materials, off season fruits, etc. With appropriate structures and plant environment control measures, the constraints of environment prevalent in the region can be overcome allowing almost round the year cultivation, increased productivity by 25%–100% and in certain cases even more, as well as conservation of irrigation water by 25%–50%. Plain architecture can be maintained through application of plant growth regulators. Most plant architecture traits can be directly retraced to changes in activity and/or size of the shoot apical meristem and derived meristems, namely lateral or axillary and floral meristems. The activity of these meristems is determined both by the plant's genetic programmes and environmental factors. Plant's growth and development depend on carbon nutrient status. Processes that especially require significant energy input are phase transitions and the initiation, and outgrowth of new shoot organs, such as floral transition and lateral branches or flowers, respectively. These processes are of vital importance to plant productivity and have major impact on reproductive output and thereby yield in many fruit crops. Timing of flowering depends on the environmental conditions and internal signals from the plant's genetic program to induce flowering in plants at the most favourable conditions. Most studies focused on highly predictable' factors involved in flowering like day length photoperiod and extended periods of cold. However, less predictable factors, including ambient temperature (15–30°C) are equally important. One or more of the factors such as temperature, light, CO_2 concentration, RH, access to insects and pests, etc., can be controlled to desirable limits to manipulate flowering and fruiting for early and multiple harvest under controlled environment. Commercial production of fruit plants particularly grown under open field conditions will be severely affected by biotic and abiotic factors increasing the gap between expected yield and obtained yield. Under polyhouse cultivation in tropical and subtropical areas, severe fungus infection along with sun burning of leaves in summer, faster growth of trees and poor quality of fruits is obtained. Due to high temperature, physiological disorders like, sun burning, sun scaled, flower and fruit abscission, seed discolouration, etc. will be more pronounced (Figure 17.1). The challenge to reduce the gap between the expected yields and observed yields can be minimised by protecting the valuable fruit crops through greater use of greenhouse technology and adoption of high-tech horticulture for judicious management of natural resources (Prakash and Singh 2006).

FIGURE 17.1 Leaf burning during summer under polyhouse cultivation due to high temperature.

17.2.1 Nutrient Management

Fertiliser application may be required depending on the location of the orchard and the variety (da Silva et al. 2007). Use of balanced applications of macro- and micronutrients helps in better crop production. Papaya plants are dependent on mycorrhiza for their nutrition and benefit greatly from soil mulching and appropriate drainage that facilitate biotic interactions in the rhizosphere and water and nutrient uptake, especially phosphorus and nitrogen (Jimenez et al. 2014). The nutrient requirements of papaya plants are high. Mineral nutrients are taken up by plants grown at full sunlight as macronutrients, namely K > N > Ca > P > S > Mg and micronutrients like, Cl > Fe > Mn > Zn > B > Cu > Mo. Nitrogen, phosphorus, and potassium, very important nutrients for metabolism and extracted in high amounts. A ton of fresh harvested fruits contains 1770, 200, and 2120 g of each of these nutrients, respectively (Jimenez et al. 2014). However, under polyhouse cultivation minimum use of nitrogenous fertiliser is recommended to avoid fast growth of plants (Morales-Payan and Stall 2005). Soil application of paclobutrazol, a growth retardant, at 1000 mg/L resulted in reduced overall height and reduced height at which first flowers bud; it did not affect the start of production or yield (Rodriguez and Galan 1995).

17.3 COMPONENTS

Protected cultivation technology has two major components. One is the infrastructure involving frames, cladding materials, irrigation system, tools, implements and other engineering inputs. These inputs ensure optimal light, air temperature, water and plant growth requirements. This optimal aspect of climatic parameters involves simple to most advanced engineering inputs such as automation, etc. to regulate several parameters such as ventilation which is one of the most important components in a successful greenhouse production. In the absence of proper ventilation, greenhouses and their plants can become prone to problems like high temperatures, development of pathogens etc. Air circulation in protected structures is a must which is done by providing ventilation devices. Importance of cladding materials in protected cultivation can hardly be overemphasised. Their quality and cost are important besides certification. Micro irrigation and fertigation involve, a lot of science and technology, demanding research for continuous improvement. This aspect of the protected structure provides optimal conditions to plants to grow normally. The advance engineering tools are assisting various operations in more precised way and ultimately contributing to harvest the quantity and quality fruits. Polyhouse and net houses are used for cultivation of this crop as given below.

17.3.1 Polyhouse

Ideal features of polyethylene have increased the use of polyhouses in place of glasshouse throughout the globe. It has not only reduced the initial cost but also increased the popularity of greenhouse by simplifying the installation technology. Generally, two types of polyhouses are used in hilly and in plain regions where simple bamboo or GI-pipe-based polyhouses are used (Singh et al. 2000).

17.3.2 Multipurpose Nets

Shade nets are used to reduce the adverse effect of scorching sun and heavy rain. Shade houses are becoming popular for growing crops and nursery during summer season. Net houses are used in high rainfall regions. Roof of the structure is covered with suitable cladding material mostly HDPE, which does not absorb moisture. Sides are made of wire mesh of different gauges; 25%–90% shade depending upon requirements. Such structures are popular in northeastern region of the country. Insect proof nets are effective to reduce the incidence of pest and viral diseases in crops. These nets are used like mosquito net around the crops, having 40–50 mesh size without colours. If we are using the green and red coloured net houses; the problems of fungus infection (Figure 17.2), poor quality of fruits (smaller and deformed) and lanky growth of trees results.

17.4 IRRIGATION

Plants depend on water within the protected structure with no rainfall to supply it naturally and the added warmth to make the soil dry out more quickly. However, near zero wind velocity under many protected environments such as polyhouses, the water requirement of many crops decreases significantly (Rana and Sah 2010). Traditional watering-cans are, of course, the simplest solution to the problem. They are particularly useful in providing very specific, targeted watering for individual plants, particularly if they are being grown in pots or containers.

17.5 TYPES OF IRRIGATION SYSTEMS

A plant nursery requires an irrigation system to water plants effectively and simultaneously. Nurserymen should think about which watering or irrigation system best suits their nursery type and size.

FIGURE 17.2 *Cercospora* leaf spot under green net house.

17.5.1 Overhead Sprinkler Irrigation Systems

Nurserymen using overhead sprinklers typically have two options. The first option is rotary sprinkler heads which contains a rotating nozzle that sends a torrent of water over plants. The second option is stationary sprinkler heads those send a rapid flow of water against a plate. The impact disrupts the steady stream of water and turns it into a continuous spray that waters plants. Although overhead sprinkler systems are the most common option in nurseries, however, they are not very efficient. They require high pressure pumps that consume large quantities of energy and also waste about 80% of the water emitted (Dubey and Singh 2012).

17.5.2 Micro Irrigation Systems

Unlike overhead sprinklers, micro-irrigation systems are highly efficient and can function using low pressure. However, soil, algae and chemical fertilisers can clog emitters for which various types of filters are provided. Three types of micro-irrigation systems are used in nurseries. One type of micro irrigation, known as the capillary mat system, uses tubes that carry water into a mat. The mat becomes saturated with water, providing containers sitting on top of the mat with a supply of water to soak up through plant root systems. Although capillary mat irrigation uses 60% less water than conventional overhead sprinkler systems, they can cause salt accumulation in the soil over long periods of time. The second type of micro-irrigation system is known as a micro-sprayer, micro-sprinkler or spray stake system. Considered one of the most efficient nursery irrigation systems, micro-sprayers use a tube to carry water directly into the soil from a water source. Not only does this eliminate water waste that is deflected off broad plant leaves, micro-sprayers carry water directly to the plant's root system. Although micro-sprayers cost more than overhead sprinklers when installed in small plants, they operate efficiently in larger plants with more foliage and heavier canopies. The third type of micro-irrigation is known as the spaghetti tube system. This nursery irrigation method uses narrow tubes to bring water into the plant container. A miniature weight at one end of the tube ensures that it stays in the container. Water travels from one pore to another, through a capillary system. Consequently, gardeners must use a high-quality, uniform soil for maximum efficiency. When using the spaghetti tube system, gardeners should keep soil moist at all times because dry soil will lead to poor water distribution (Dubey and Singh 2012).

17.5.3 Capillary Sand Beds

Unlike sprinkler and micro-irrigation systems, capillary sand beds do not involve any electricity. Containing wood panels, a plastic liner, sand, a small water reservoir, drainage pipe and valve, capillary sand beds are built to slant slightly, allowing water released into the raised end to slowly travel to the lower end. Providing an even and continuous water supply, capillary sand beds involve less maintenance. Plants grow evenly, relying less on fertiliser and pesticide. However, capillary sand beds attract weeds and also have high installation costs.

17.5.4 Overhead Systems

In overhead misting systems, sprinklers are installed under the roof framework of polyhouse, and this 'rain' water down pours onto plants. This type of irrigation system is easy to automate and produces high humidity. This high humidity allows protecting crops against frost damage. For the best coverage, space overhead sprinklers to around 50%–60% of the wetting diameter of the sprinkler.

17.5.5 Bench Misting

Bench misting uses a central line of sprinklers or hoops placed at or just above the level of plants. Bench misting requires plants to be placed on raised benches and these must be made of materials that are impervious to water, such as a metal. This allows having just a single misting bench in the polyhouse and a different watering system in the rest of the polyhouse.

17.5.6 Drip Irrigation

Drip irrigation is one of the most efficient methods of watering, typically operating at 90% efficiency. Runoff and evaporation are at a reduced rate compared to other irrigation systems such as sprinklers. In drip irrigation, tubes that have emitters run alongside the plants receiving irrigation. The water leaves from the tubes through the emitters by slowly dripping into the soil at the root zone. This method of irrigation minimises leaf, fruit and stem contact with water resulting in reduced plant disease. It reduces weed growth by keeping the area between plants dry. Drip watering is excellent way to water in polyhouse as it keeps the humidity low leading to less pest and disease problems. Water is directed to exactly where it is needed either with an individual dripper, especially for pot grown plants, or inserted into a pipe for beds (Dubey and Singh 2012). Protected cultivation improves the water productivity due to the evapotranspiration (ET) reduction and larger outputs of protected growing (Stanghellini 1993). Drip irrigation also increases the water productivity, relative to conventional irrigation in greenhouses (Castilla 1994), far from the 65 kg/m^3 obtained in sophisticated greenhouses, with soilless culture and very long cycles (Stanghellini 1993).

17.5.7 Hand Watering

Hand watering is the most basic method of irrigation for a greenhouse or polyhouse. Watering cans or hoses with nozzles allow tailoring watering to the needs of individual plants. This is also the most labour-intensive method for plant irrigation and may not be effective for large scale cultivation or nursery raising under polyhouses.

17.6 DISEASE MANAGEMENT IN POLYHOUSES

Papaya is very sensitive to viral diseases under protected environment. It often spreads in the plantation by insect vectors such as whitefly, thrips and aphids.

FIGURE 17.3 Nursery raising under polyhouse.

The damage caused by the virus is usually much greater than the mechanical injury caused by the insect vector. Bacterial diseases are less frequent but under high moisture and poor irrigated condition may cause huge damage, namely *Erwinia*, *Xanthomonas*, etc. in papaya.

Fungal diseases constitute one of the biggest groups of foliar pathogens causing immense damage under protected environment. It was found that the incidence and severity of diseases vary considerably under protected environment when compared to the open field. As observed in papaya, *Phytophthora, Pythium, Fusarium, Alternaria* and *Colletotricum* were observed to be of higher significance under polyhouse conditions (Dubey and Singh 2012). This may be attributed to the fact that temperature and humidity are nearly balanced inside protected structure, even when outside field temperature is comparatively low and so on. Survival of pathogen is also enhanced inside polyhouse due to availability of host because of longer growing season (Figure 17.3).

17.6.1 Management Strategy

Proper field sanitation is one of the most important management strategies. Since, once the inoculums build up inside polyhouses it is very difficult to manage it. Incidence of different diseases was managed by providing ample spacing between plants for proper air movement and reducing of leaf wetness by avoiding overhead irrigation. Balance use of chemicals with least toxicity is recommended, especially for polyhouse cultivation. Scout the fields for the first occurrence of virus disease. Pull up and destroy infected plants, but only after spraying them thoroughly with an insecticide to kill any insects they may be harbouring. Use reflective mulches to repel insects, thereby reducing the rate of spread of insect borne viruses. Monitor vector population early in the season and apply insecticide treatments when needed. Minimise plant handling to reduce the amount of virus spread mechanically. Composts have long been known to improve soil fertility and plant disease management (Dubey and Singh 2012).

Compost improve the ability of plants to resist against disease caused by root pathogen like, *Fusarium, Phytophthora, Pythium, Rhizoctonia*, etc. and foliar pathogens like, *Pseudomonas, Colletotrichum, Xanthomonas*, etc. Among soil-borne root pathogens, suppression of *Fusarium* using composts has been reported earlier (Punja et al. 2002). The severity of various diseases caused by *Fusarium* has been reduced from 20% to 90% using compost amendments. Composts have also been used successfully for suppression of *Phytophthora* crown and root rots of nursery, fruit crops produced in container media (Aryantha et al. 2000) and field soils (Downer et al. 2001). Addition of compost serves two possible purposes significant for the biological control of *Phytophthora cinnamomi* where it provides a substrate for the growth of fungal antagonists and creates an environment that promotes enzyme activity (Downer et al. 2001). The suppression of diseases caused by *Pythium* spp. has been well-documented. The severity of diseases caused by the fungus, *Pythium* was reduced by 30%–70% when growing media were amended with various compost products.

17.7 PROSPERITY

Infection of virus and viral diseases are becoming serious threats to the papaya industry in India. Papaya is a shy crop to application of pesticides. The net-house-based production system altered with the need-based use of polyethylene sheet offers immense scope to enhance papaya fruit production with assured yield and fruit quality. The intervention of net house will be able to reduce the movement of vectors of the viral diseases and pests like fruit flies. The papaya crop is damaged due to frost in the northern part of India which can be protected under greenhouse cultivation. The protected environment will reduce abnormalities in bisexual flowers of the gynodioecious varieties or hybrids which leads the misshapen fruits with poor marketability. The papaya fruit yield is around 10 kg/plant which contributes to 40 t/ha productivity. The favourable growing conditions are able to enhance the fruiting zone from 0.60 to 1.5 m which facilitates proper space for fruit set, growth and development. The enhanced fruiting zones also offer retention of 40–65 fruits/plant with uniform size and quality. Fruit yields of 45–60 kg/plant can be obtained under protected cultivation, which may lead to higher productivity in a range of 180–240 t/ha (Prakash and Singh 2006).

17.8 FUTURE PROSPECTS

The input use and their effectiveness under protected environment have to be standardised. The pollinator and pollination is vital for the fruit set and retention which needs further study under protected conditions. The biotic stresses like mites and nematodes are major concern in protected production system, and their dynamics have to be studied. The economic feasibility of the protected production system has to be investigated considering the large number of stakeholders as small and marginal farmers in the country. The challenges to the horticulturists and breeders are not going to end up here due to rising demand of trait specific hybrids or cultivars of the fruit crops under protected cultivation. This fruit crop is able to complete its life cycle in short period, getting damaged with insects as pests or vectors and able to utilise the vertical space of the poly/green houses shall be fit for the protected cultivation in tropical and subtropical regions of India.

18 Descriptors

Papaya shows considerable phenotypic variation in morphological and horticultural traits that can be utilised in its genetic improvement. On account of high degree of variability in economic characters, several selections from the existing population have been made in different papaya growing areas of the world. The objective of this chapter was to provide information for collection, documentation and characterisation of the papaya germplasm. The wide diversity can be utilised for the selection of promising parents in hybrid variety, inbred line development and estimating the potential of genetic gain in a breeding programme. There is also need for proper conservation of the different accessions reported as they could serve as raw material for the genetic improvement of different characters of the crop through recurrent selection after hybridisation. Morphological and agronomical characterisation is usually an affordable way to estimate the genetic diversity and is essential for genotype discrimination and identification of duplicates. The morphological characteristics were recorded and data from accessions submitted to principal component and cluster analysis. The collected papaya germplasms were characterised in the field using IBPGR (International Bureau of Plant Genetic Resources) morphological descriptors based on fruit, flower, stem and leaf characteristics. The lists include accession data, collection data, site and plant data. Additional data to be collected for further characterisation and evaluation of papaya are listed (site data, plant data, stress susceptibility, pest and disease susceptibility, alloenzyme composition, cytological characters and identified genes, and notes) below (IBPGR 1988).

18.1 CHARACTERISATION AND PRELIMINARY EVALUATION

1. Site Data
    - 1.1 Country of characterisation and preliminary evaluation
    - 1.2 Site (research institute)
    - 1.3 Name of person(s) in charge of characterisation
    - 1.4 Sowing date
    - 1.5 Percentage germination
    - 1.6 Number of days to 50% germination
    - 1.7 Transplanting date
    - 1.8 Tree site in the field

Give block, strip and/or row numbers as applicable

- 1.9 Spacing in the field
- 1.10 Soil type
    1. Clay
    2. Clay-silt

3. Silt
4. Loam
5. Silt-sand
6. Sand
7. Highly organic

1.11 Watering
1. Irrigated
2. Rained

1.12 First harvest date
1.13 Last harvest date in first production year
1.14 Last harvest date

2. Plant Data

2.1 Vegetative

2.1.1 Tree habit
1. Single stem
2. Multiple stems

2.1.2 Number of nodes to first flower
2.1.3 Length of middle internode on tree (cm)

Mean of 5 measurements

2.1.4 Stem colour (adult trees)
1. Greenish or light grey
2. Greyish brown
3. Green and shades of red-purple (pink)
4. Red-purple (pink)
5. Other (specify)

2.1.5 Stem pigmentation
1. Only or mostly basal
2. Only or mostly lower
3. Only or mostly median
4. Only or mostly indiscriminate

2.1.6 Colour of mature leaf petiole
1. Pale green
2. Normal green
3. Dark green
4. Green and shades of red-purple
5. Red-purple
6. Other (specify in the notes descriptor, 8)

2.1.7 Leaf shape

2.2 Inflorescence and fruit

2.2.1 Type of tree hermaphroditism
1. Staminate flowers and a few hermaphrodite flowers
2. A few staminate flowers and many hermaphrodite flowers
3. A few staminate flowers, many hermaphrodite flowers and a few pistillate flowers

4. Hermaphrodite flowers only
5. Hermaphrodite flowers and a few pistillate flowers
6. A few hermaphrodite flowers and many pistillate flowers

2.2.2 Type of flowering
1. Flowers solitary (singly borne)
2. Inflorescences
3. Both

2.2.3 Colour of inflorescence stalk
1. Greenish
2. Purplish/pinkish
3. Dark red-purple/pink

2.2.4 Predominant inflorescence size
1. Small
2. Intermediate
3. Large

2.2.5 Flower size (specify sex)

Observed on completely developed open flowers

1. Generally small
2. Generally intermediate
3. Generally large

2.2.6 Corolla tube colour of male flower

Observed on completely developed open flowers

1. White
2. White yellow (cream)
3. Yellow
4. Deep yellow to orange
5. Greenish
6. Dark green
7. Yellow/green and red-purple shades
8. Red purplish (pinkish)
9. Dark red-purple (pink)
10. Other (specify in the notes descriptor, 8)

2.2.7 Corolla lobes colour of male flowers

Observed on completely developed open flowers

1. White
2. White yellow (cream)
3. Yellow
4. Deep yellow to orange
5. Greenish
6. Dark green

7. Yellow/green and red-purple shades
8. Red purplish (pinkish)
9. Dark red-purple (pink)
10. Other (specify in the notes descriptor, 8)

2.2.8 Colour of female flower

Observed on completely developed open flowers

1. White
2. White yellow (cream)
3. Yellow
4. Deep yellow to orange
5. Greenish
6. Dark green
7. Yellow/green and red-purple shades
8. Red purplish (pinkish)
9. Dark red-purple (pink)
10. Other (specify in the notes descriptor, 8)

2.2.9 Colour of hermaphrodite flower

Observed on completely developed open flowers

1. White
2. White yellow (cream)
3. Yellow
4. Deep yellow to orange
5. Greenish
6. Dark green
7. Yellow/green and red-purple shades
8. Red purplish (pinkish)
9. Dark red-purple (pink)
10. Other (specify in the notes descriptor, 8)

2.2.10 Fruit shape (fruits from hermaphrodite flowers)

Scored at full development

1. Globular
2. Round
3. High round
4. Elliptic
5. Oval
6. Oblong
7. Oblong-ellipsoid
8. Oblong-blocky
9. Elongate
10. Lengthened cylindrical

11. Pear shaped (pyriform)
12. Club
13. Blossom end tapered
14. Acron (heart shaped)
15. Reniform
16. Turbinate inferior
17. Plum shaped
18. Other (specify/describe)

2.2.11 Fruit shape (fruits from female flowers)

Scored at full development

1. Globular
2. Round
3. High round
4. Elliptic
5. Oval
6. Oblong
7. Oblong-ellipsoid
8. Oblong-blocky
9. Elongate
10. Lengthened cylindrical
11. Pear shaped (Pyriform)
12. Club
13. Blossom end tapered
14. Acron (heart shaped)
15. Reniform
16. Turbinate inferior
17. Plum shaped
18. Other (specify/describe)

2.2.12 Fruit skin colour

Overall colour of the skin of ripe fruits

1. Yellow
2. Deep yellow to orange
3. Red/purple
4. Yellowish green
5. Green
6. Other (specify in the notes descriptor, 8)

2.2.13 Fruit flesh colour

Observe on ripe fruits

1. Light yellow
2. Bright yellow
3. Deep yellow to orange

4. Reddish orange
5. Scarlet
6. Other (specify in the notes descriptor, 8)

2.2.14 Tree fruit productivity (kg per annum)

1. Low (approximately 10 kg)
2. Intermediate (approximately 50 kg)
3. High (approximately 80 kg)
4. Extremely high (approximately 110 kg)

2.3 Seed

2.3.1 Seed colour

1. Generally tan
2. Generally grey-yellow
3. Generally grey
4. Generally brown black
5. Generally black
6. Variable

2.3.2 Seed germinating in ripe fruit

0 Absent
+ Present

2.3.3 100-seed weight (g)

18.2 FURTHER CHARACTERISATION AND EVALUATION

3. Site Data

3.1 Countries of further characterisation and evaluation
3.2 Site (research institute)
3.3 Name of person(s) in charge of characterisation
3.4 Sowing date
3.5 Percent germination
3.6 Number of days to 50% germination
3.7 Transplanting date
3.8 Tree site in the field

Give block, strip and/or row numbers as applicable

3.9 Spacing in the field
3.10 Soil type

1. Clay
2. Clay-silt
3. Silt
4. Loam
5. Silt-sand
6. Sand
7. Highly organic

3.11 Watering
1. Irrigated
2. Rainfed
3.12 First harvest date
3.13 Last harvest date in first production year
3.14 Last harvest date

4. Plant Data

Unless otherwise noted, descriptors should be evaluated in the first year of production

4.1 Vegetative
4.1.1 Tree height

Measured from the ground to apical meristem at first harvest

1. Short (<1 m)
2. Intermediate
3. Tall (>1 m)

4.1.2 Tree diameter (mm)

To be measured 10 cm above the ground

4.1.3 Height at first fruit
1. Low bearing (<1.0 m)
2. Intermediate
3. High bearing (>1.5 m)

4.1.4 Length of mature leaf petiole (cm)

Average of five middle leaves

4.1.5 Length of mature leaf (cm)

Average of same five leaves, and measured from base of middle leaflet midrib to tip

4.1.6 Width of mature leaf (cm)

Average of the same five leaves, and measured at maximum breadth

4.1.7 General shape of mature leaf teeth
1. Straight
2. Convex
3. Concave
4. Other (specify in the notes descriptor, 8)

5. Inflorescence and Fruit
5.1 Density of inflorescences on trunk

Observe several trees before scoring

1. Sparse (few inflorescences)
2. Intermediate
3. Dense (many inflorescences)

5.2 Inflorescence density

Density of flowers within free inflorescences

1. Sparse (few inflorescences)
2. Intermediate
3. Dense (many inflorescences)

5.3 Length of inflorescence main axis (cm)

Average of five basal (old) inflorescences

5.4 Corolla length of male flowers (cm)

1. Generally short
2. Generally long

5.5 Corolla length of hermaphrodite flowers (cm)

Observe several hermaphrodite flowers before scoring

1. Generally short
2. Generally long

5.6 Corolla length of female flowers (cm)

Observe several female flowers before scoring

1. Generally short
2. Generally long

5.7 Sex change of flowers during growth: male to hermaphrodite

0 No
+ Yes

5.8 Sex change of flowers during growth: hermaphrodite to male

0 No
+ Yes

5.9 Sex change of flowers during growth: hermaphrodite to female

0 No
+ Yes

5.10 Number of flowers per node
5.11 Number of fruits per node
5.12 Uniformity of fruit distribution

0 Not uniform
+ Uniform

5.13 Number of fruits on trunk

An average of five plants should be taken from a 1–2 year fruiting season

5.14 Length of peduncle (cm)

Average of five plants

5.15 Skin colour of immature fruits
1. Yellow
2. Light green
3. Green
4. Other (specify in the notes descriptor, 8)

5.16 Stalk end fruit shape
1. Depressed
2. Flattened
3. Inflated
4. Pointed

5.17 Size of blossom end scar
1. Small (<0.5 cm)
2. Intermediate
3. Large (>1.0 cm)

5.18 Fruit skin texture when ripe
1. Smooth
2. Intermediate
3. Rough (ridged)

5.19 Ridging on fruit surface
1. Superficial (low depression)
2. Intermediate (moderate depression)
3. Deep (usually five distinct ridges)

5.20 Fruit weight (g)

Average of five fruits

5.21 Fruit length (cm)

To be measured from base of calyx to tip of fruit. Average of five fruits

5.22 Fruit diameter (cm)

To be measured at the broadest part. Average of five fruits

5.23 Shape of central cavity

To be determined as fruit cut open (cross-section) at maximum diameter

1. Irregular
2. Round
3. Angular
4. Slightly star shaped
5. Star shaped
6. Other (specify in the notes descriptor, 8)

5.24 Central cavity diameter (cm)

Measured at maximum diameter. Average of five fruits

5.25 Thickness of fruit skin

1. Thin
2. Intermediate
3. Thick

5.26 Flesh aroma

1. Mild
2. Intermediate
3. Strong

5.27 Flesh density

1- Very low (spongy)
3- Low (crumbly)
5- Intermediate
7- Dense (crisp)
9- Very dense (firm)

5.28 Flesh fibrousness

0 Absent
+ Present

5.29 Placental tissue

2- Little
5- Intermediate
7- Much

5.30 Eating quality (dessert)

A combined assessment of flavour, sweetness and aroma when ripe

3- Poor
5- Intermediate
7- Good
9- Excellent

5.31 Seed
5.31.1 Fresh weight of seeds per fruit (g)

Average of five fruits

5.31.2 Seed surface lustre

3- Generally dull
5- Generally intermediate
7- Generally glossy

5.31.3 Seed shape
1. Generally round
2. Generally spherical or ovoid
3. Other (specify in the notes descriptor, 8)

5.31.4 Seed surface type
1. Generally translucent
2. Generally opaque

5.31.5 Seed mucilage
1- Almost absent
3- Small
5- Intermediate
7- Large

5.32 Tree yield data

Specify tree age

5.32.1 Leaf yield per tree (kg)

Total fresh weight of leaves harvested over one season (or year)

5.32.2 Total dry papain yield per tree (kg)

Total weight from leaves, trunks and unripe fruits over one season (or year)

5.32.3 Total number of harvested fruits per season (or year)
5.32.4 Total weight of harvested fruits per season (or year) (kg)
5.32.5 Total number of harvests per season (or year)
5.32.6 Total dry seed yield per season (or year) (kg)

5.33 Chemical data
5.33.1 Percentage leaf dry matter (%)

5.33.2 Leaf protein content (%)

Measured as percentage of fresh weight

5.33.3 Leaf mineral content (%)

Measured as percentage of leaf dry matter

5.33.4 Papain oxidation
5.33.5 Refractometer reading of fruit juice
5.33.6 Total soluble solids of fruit flesh (%)

Expressed as percentage of fresh weight of mature fruit flesh

5.33.7 Percentage of ash in fruit flesh (%)
5.33.8 Percentage of acids in fruit flesh (%)
5.33.9 Percentage of protein in fruit flesh (%)
5.33.10 Percentage of total sugars in fruit flesh (%)
5.33.11 Percentage of fat in fruit flesh (%)
5.33.12 Percentage of fibre in fruit flesh (%)
5.33.13 Percentage oil content of seed (%)

6. Stress Susceptibility

To be scored on a 1–9 scale where
1- Very low susceptibility
3- Low susceptibility
5- Medium susceptibility
7- High susceptibility
9- Very high susceptibility

6.1 Reaction to low temperature (frost susceptibility)
6.2 Reaction to drought
6.3 Reaction to high soil moisture (water-logging)

7. Pest and Disease Susceptibility

To be scored on a 1–9 scale where
1- Very low susceptibility
3- Low susceptibility
5- Medium susceptibility
7- High susceptibility
9- Very high susceptibility

7.1 Pests
7.1.1 *Aphis gossypii* Glover cotton or melon aphid
7.1.2 *Aphis craccivora* Koch Cowpea aphid

7.1.3 *Aphis middletonii* Erigeron root aphid Thomas
7.1.4 *Aphis spiraecola*
7.1.5 *Heteromyzus lactucae* L. Sonchus aphid (=*Amphorophora sonchi* Oestlund)
7.1.6 *Macrosiphum euphorbiae* Thomas potato aphid
7.1.7 *Neomyzus circumflexus* Buckton crescent-marked lily aphid
7.1.8 *Myzus persicae* Sulzer Green peach aphid
7.1.9 *Rhopalosiphum maidis* Fitch corn leaf aphid
7.1.10 *Exillis lepidus* Jordan fungus weevil
7.1.11 *Rhabdoscelus obscurus* Boisduval New Guinea sugarcane weevil
7.1.12 *Ceratitis capitata* Weidemann mediterranean fruit fly
7.1.13 *Dacus cucurbitae* Coquillet melon fly
7.1.14 *Dacus tryoni* Froggatt Queensland fruit fly
7.1.15 *Dacus cucuminis* French cucumber fly
7.1.16 *Dacus neohumeralis* Lesser Queensland fruit fly Hardy
7.1.17 *Dacus dorsalis* Hendel oriental fruit fly
7.1.18 *Chrysoma megacephala* oriental blowfly Fabricius
7.1.19 *Neoexaireta spinigera* blue soldier fly Weidemann
7.1.20 *Volucella obesa* Fabricius green syrphid fly
7.1.21 *Toxotrypana curvicauda* papaya fruit fly
7.1.22 *Nezara viridula* L. southern green stink bug
7.1.23 *Amblypelta lutescens* distant banana spotting bug
7.1.24 *Empoasca solana* Delong
7.1.25 *Empoasca papaya*
7.1.26 *Agrotis ipsilon* Aufnagel
7.1.27 *Heliothis hawaiiensis* quaintance
7.1.28 *Heliothis zea* Boddie native budworm
7.1.29 *Heliothis punctigera* wallengren corn earworm
7.1.30 *Othreis fallonia* Clerck fruit-sucking moth
7.1.31 *Othreis materna* L. fruit-sucking moth
7.1.32 *Eudocima salaminia* Cramer fruit-sucking moth
7.1.33 *Dichocrocis punctiferalis* Guinee yellow peach moth
7.1.34 *Cryptoblades aliena*
7.1.35 *Aspidiotus destructor* coconut scale Signoret
7.1.36 *Coccus elongatus* long brown soft scale Signoret
7.1.37 *Coccus hesperidum* L. brown soft scale
7.1.38 *Howardia biclavis* mining scale comstock
7.1.39 *Pseudoparlatoria ostriata*
7.1.40 *Pseudococcus obscurus* obscure mealybug Essig
7.1.41 *Thrips tabaci* Lindeman onion thrips
7.1.42 *Trialeurodes* greenhouse whitefly vaporariorum Westwood
7.1.43 *Bemisia* spp. Whitefly
7.1.44 *Brevipalpus phoenicis* red and black flat mites Geijskes
7.1.45 *Eutetranychus banksii* Texas citrus mite McGregor
7.1.46 *Panonychus citri* citrus red mite McGregor

7.1.47 *Tetranychus cinnabarinus* carmine mite Boisduval
7.1.48 *Tetranychus urticae* Koch two spotted mite
7.1.49 *Hemitarsoneumus latus* broad mite Banks
7.1.50 *Tuckerella omata* Tucker ten-tailed tuckerellid
7.1.51 *Tuckerella pavoniformis* twelve-tailed McGregor tuckerellid
7.1.52 *Tenuipalpus bioculatus*
7.1.53 *Meloidogyne* spp root-knot nematodes
7.1.54 Other (Specify in the notes descriptor, 8)

7.2 Fungi
7.2.1 *Altemaria* spp. Altemaria rot
7.2.2 *Ascochyta caricae* papaya leaf and fruit spot
7.2.3 *Ascochyta* spp. *Ascochyta* rot
7.2.4 *Asperisporium caricae* papaya leaf blight
7.2.5 *Cercospora papayae* black spot
7.2.6 *Cladosporium* spp. blossom-end rot
7.2.7 *Colletotrichum gloeosporioides* Anthracnose
7.2.8 *Corynespora cassiicola* leaf spot
7.2.9 *Fusarium* spp. stem end rot
7.2.10 *Glomerella cingulata* fruit rot
7.2.11 *Oidium caricae* powdery mildew
7.2.12 *Phytophthora parasitica*/Phytophthora fruit rot *P. palmivora*
7.2.13 *Pythium* spp./collar and root rot *Phytophthora parasitica*/*P. palmivora*
7.2.14 *Rhizoctonia* spp. damping-off
7.2.15 Other (specify in the notes descriptor, 8)

7.3 Bacteria

Specify in the notes descriptor, 8

7.4 Virus and Mycoplasma
7.4.1 Papaya mosaic
7.4.2 Papaya ringspot
7.4.3 Papaya bunchy top
7.4.4 Yellow crinkle
7.4.5 Tomato big bud organism
7.4.6 Other (specify in the notes descriptor, 8)

7.5 Other Disorders
7.5.1 Freckles
7.5.2 Bumpyness
7.5.3 Dieback (unknown cause)
7.5.4 Other (specify in the notes descriptor, 8)

8. Notes: Give additional information where the descriptor state is noted.

The major limitation of morphoagronomic characterisation is that many of the traits used have polygenic inheritance and are influenced by the environment. Besides, the identification of papaya cultivars using morphological traits is not usually possible until fruit production. Papaya germplasm shows considerable phenotypic variation for many horticultural traits (Ocampo et al. 2006). The different criteria, which are used to estimate genetic diversity, include pedigree records, morphological traits and molecular markers. Plant taxonomy is traditionally dependent upon the comparative external morphological characters (Baxy 2009). However, these are environmental and developmental stage dependent. Therefore, molecular markers are preferred choice for plant identification as they are detectable in all tissues and independent of environmental changes (Tapia et al. 2005). For the development of papaya hybrids, pure lines are needed in order to avoid F_1 segregation. Classical improvement procedures to obtain papaya lines are based on the inbreeding of segregating populations and of germplasm accessions of *Carica papaya* L. This strategy results in the selection of pure lines through progeny testing, which can take up to five to six inbreeding generations (average 12 years), depending on the genetic diversity of the background. The pure lines are then identified among the progenies whose phenotypical segregation is considered to be null. Notwithstanding, this methodology is expensive, laborious, time- and space-consuming, and influenced by negative conditions in the environment.

Genetic diversity was determined using several molecular markers namely, random amplified polymorphic DNA (RAPD), inter-simple sequence repeat (ISSR), amplified fragment length polymorphism (AFLP), simple sequence repeat (SSR) markers, and so on, have been used for analysis of genetic diversity, relationships and germplasm identification of papaya (Jesus de et al. 2012), computing allelic richness and frequency, expected heterozygosity and cluster analysis. Among the molecular markers, RAPD and ISSR markers have been extensively used to study genetic diversity and relationships in papaya germplasm (Rodriguez et al. 2010; Sudha et al. 2012; Saran et al. 2015). Among these molecular markers, ISSRs and SSRs are considered to be useful and have been extensively used for the identification of species or germplasm in a wide range of plants (Ahmad et al. 2010). ISSR amplifies inter-microsatellite sequences at multiple loci throughout the genome (Li and Xia 2005) and permits the detection of polymorphism in microsatellites and inter-microsatellite loci without previous knowledge of DNA sequences. These markers can detect polymorphism in a single reaction with a good repeatability and reproducibility (Saran et al. 2015). Knowledge of genetic variability is very useful for identifying the best combinations between germplasm lines with the potential to maximise the genetic gains attained by hybridisation (Bertan et al. 2009). Although the level of genetic diversity revealed by SSR markers is sufficient to distinguish between breeding lines for varietal protection, the rather narrow genetic diversity demonstrated indicates the need to introduce new germplasm or use other techniques such as mutation and genetic engineering to provide breeding materials for the future improvement. Genetic diversity was determined using seven simple sequence repeat (SSR) markers. Although the level of genetic diversity revealed by SSR markers in this study is sufficient to distinguish between breeding lines for varietal protection, the rather narrow genetic diversity demonstrated indicates the need to introduce new germplasm or use other techniques such as mutation and genetic engineering to provide breeding materials for the future improvement of papaya (Asudi et al. 2013).

The limited number of papaya varieties available reflects the narrow genetic base of this species. The use of backcrossing as a breeding strategy can increase variability, besides allowing targeted improvements. Procedures that combine the use of molecular markers and backcrossing permit a reduction of the time required for introgression of genes of interest and appropriate recovery of the recurrent genome. Microsatellite markers have been used to characterise the effect of first-generation backcrosses of three papaya progeny, by monitoring the level of homozygosity and the parental genomic ratio (Ramos et al. 2011). Among the molecular markers currently available, microsatellites (simple sequence repeats, SSRs) are of particular importance because they show extensive polymorphism as a consequence of the occurrence of different numbers of repeated units within the SSR structure (Morgante and Olivieri 1993). SSRs have other advantages as well, such as high reproducibility, technical simplicity, low cost, high resolution power and, most important of all, codominance (Rallo et al. 2000; Oliveira et al. 2006). SSR have also been widely used for marker-assisted selection in backcross programs (Benchimol et al. 2005; Xi et al. 2008). In the preceding years, SSRs have become one of the most popular molecular markers due to the massive amount of sequences available in databases, reflecting the progress of genome research (Leal et al. 2010). In plant breeding as well as genetic analysis, this marker has had a variety of applications due to its multi-allelic nature, reproducibility, high information content, codominant inheritance, high abundance and extensive coverage of the genome (Gupta and Varshney 2000) and distribution in a non-random way (Wang et al. 2008). In the papaya genome, microsatellites are more abundant type of tandem repetition, with a density of one every 0.7 kb; however, it represents only 0.19% of the entire genome of this species (Moore and Ming 2008). Microsatellites were first reported in *C. papaya* by Oliveira et al. (2008) and Eustice et al. (2008). Reliable and highly polymorphic SSR markers have been used mainly for genetic map construction (Chen et al. 2007), sexual differentiation (Santos et al. 2003) and to access genetic diversity (Ocampo Perez et al. 2007).

The use of co-dominant markers, such as SSRs, in MAS allows the early identification of plants with high levels of homozygosity in segregating progenies and germplasm accessions. Procedures involving the use of molecular markers and the indirect selection of homozygous plants can reduce this time considerably. The use of molecular markers, particularly those used in marker-assisted selection (MAS) has allowed important progress in terms of crop improvement. Marker-assisted selection (MAS) microsatellite marker is a quick and effective procedure for the development of new papaya lines (de Oliveira et al. 2010).

Glossary

Abortive: Only partially developed, such as incomplete seed or frost-nipped bud.

Abscission: The falling of leaf, twig-tip, and so on, from a clean-cut scar by a self-healing wound.

Acidic soil: Soil with a pH measure below 7 and also called sour soil.

Acre: A measure of land totalling 43,560 square feet. A square acre is 208.75 feet on each side.

Aeration: Introduction of air to compacted soil by mechanically removing plugs of top soil. Aeration helps oxygen, water, fertiliser and organic matter to reach roots.

Air layering: A specialised method of plant propagation accomplished by cutting into the bark of the plant to induce new roots formation. It is a process of producing a new plant by forming soil around a stem.

Alkaline soil: A soil with a pH higher than 7.0 is an alkaline soil. pH is a measure of the amount of lime (calcium) contained in soil.

Amendments: Organic or mineral materials, such as peatmoss, compost, perlite, and so on, are used to improve the soil health.

Androecium: Stamens of a flower, as a unit.

Andromonoecious: Refers to a plant species in which male and bisexual flowers are produced on the same plant.

Anther: The pollen-bearing part of a stamen, borne at the top of a filament or sessile.

Anthracnose: A fungal disease that causes spots on foliage or fruit and often death of plant.

Aphid: A small green or white insect that sucks juices from plant parts.

Articulated laticifers: A laticifer is a type of elongated secretary cell composed of a series of cells joined together found in the leaves and/or stems of plants that produce latex and rubber as secondary metabolites.

Asexual reproduction: Duplicating a plant from any cell, tissue or organ of that plant.

Auxin: A chemical that stimulates plant growth.

Backcross: A hybrid of two plants crossed once again back to one parent.

Berry: Berry is a fleshy fruit produced from a single flower and contains one ovary or it is a pulpy fruit developing from a single pistil, containing one or more seeds but no true stone.

Biennial: A plant sown one year to flower or fruit the next, then dying or being discarded.

Biological control: Reducing pests by utilising other organisms.

Boron: A trace mineral needed by the body in only very small amounts. It assists in the proper absorption of calcium, magnesium and vitamin D.

Carica papaya: Papaya is a giant herbaceous plant resembling a tree but not woody in the Caricaceae family that originated in Central America and is now grown in tropical areas worldwide for its large, sweet and melon-like fruits.

The name 'papaya' also refers to the fruit of other Carica species, including *C. pubescens* and *C. stipulata*, and their various hybrids.

Carpel: A single pistil in a female flower part containing several pistils.

Certification: Seed or stock verified to be true or lacking viruses.

Certified seed: Progeny of registered seed stock is the final stage in the expansion programme and is certified with a metal seal and blue tag.

Chelating agent: An organic compound that keeps metals in water from combining and also known as sequestering agents, which prevent metal buildup that causes staining.

Chilling injury: Damage to certain crops, namely, papaya, banana, cucumber and sweet potato, which results from exposure to cold but above-freezing temperatures.

Chlorosis: The yellowing of plant tissue due to nutrient deficiencies or disease.

Chromosomes: Small structures in the nucleus of a cell that carry the genes. They appear as thread or rod-shaped structures during metaphase. Each species has a characteristic number of chromosomes.

Climacteric period: The period in the development of some plant parts that involves a series of biochemical changes associated with the natural respiratory rise and autocatalytic production of ethylene. The climacteric period consists of the pre-climacteric, pre-climacteric minimum, climacteric rise, climacteric peak and post-climacteric phases.

Clone: A group of individuals of common ancestry, which have been propagated vegetatively (asexually), usually by cutting or natural multiplication of bulbs or tubers.

Cold storage: A type of insulated storage utilising mechanical refrigeration to maintain a stable cold temperature for long-term storage.

Commercial hybrid: Refers to the first-generation F_1 of the hybrid planted for any purpose.

Complete fertiliser: A fertiliser that contains nitrogen, phosphorus and potassium.

Complete flower: Flower composed of a short axis or receptacle from which arise four sets of floral parts-sepals, petals, stamens and pistils.

Compost: Humus made by decomposing vegetative matter in a compost bin or pile.

Cover crop: A crop that is planted in the absence of the normal crop to control weeds and add humus to the soil when it is ploughed in prior to regular planting.

Cross-pollination: The process in which pollen is transferred from an anther (the upper part of the stamen in which pollen is produced) of one flower to the stigma (the pollen-receiving site of the pistil) of a second flower of a different cultivar.

Cultivar: A plant derived from a cultivated variety that has originated and persisted under cultivation, not necessarily referable to a botanical species, and of botanical or horticultural importance, requiring a name.

Cytoplasm: Main contents of a cell in which the nucleus and other bodies are located.

Damping off: Decay of young seedlings at the ground level following a fungal attack.

Dioecious: This species having unisexual flowers and each sex confined to a separate plant or having staminate and pistillate flowers on different plants of the same species, for example, spinach, hops, hemp, date palm, papaya, pointed gourd, little gourd, asparagus, and so on.

Diploid: Organism or cell with two sets of chromosomes.

Disbudding: The removal of unwanted vegetative or flower buds.

Diseased plant: A plant that is abnormal because of a disease-causing organism or virus.

Drip irrigation: The application of small quantities of water directly to the root zone through various types of delivery systems on a daily basis.

Drip line: The circle that would exist if you draw a line below the tips of the outer most branches of a tree or plant.

Dwarf: A genetically one-fourth or less smaller plant as compared to normal size due to lesser intermodal space.

Emasculation: Removal of stamens before they burst and shed their pollen.

Endemic: Native or local to an area.

Endocarp: The inner area of the fruit wall.

Ethylene: Naturally occurring gas responsible for fruit ripening.

Etiolation: Stretching of a plant and loss of colour due to a lack of needed light.

Exocarp: The outer skin like region of the fruit wall.

F_1: Generation that arises from a given crossing and known as the filial generation.

F_2: Generation produced by selfing the F_1 known as the second filial generation.

F_1 hybrid: A first generation of a cross of different plants.

Feminisation: The hormonally induced development of female sexual characteristics.

Fertile: Capable of producing seed.

Fertiliser: Organic or inorganic plant food used to amend the soil in order to improve the quality or quantity of plant growth.

Field inspection: An official inspection of seed fields conducted by the official of a certification agency or his authorised agent.

Fleshy fruits: Classification of fruits that includes the berry, drupe and pome. They have a fruit wall that is soft and fleshy at maturity.

Flower: A shoot of determinate growth with modified leaves that is supported by a short stem and structure involved in the reproductive processes of plants that bear enclosed seeds in their fruits.

Foliar feeding: Fertiliser applied in the liquid form to the plants foliage in a fine spray.

Food and Agriculture Organization (FAO): An agency of the United Nations serving both developed and developing countries. It helps developing countries in transition modernise, improve agriculture, forestry and fisheries practices and also ensure good nutrition for all.

Foundation seed: Seed stock handled to most nearly maintain specific genetic identity and purity under supervised or approved production methods certified by the agency.

Fruit: An expanded, ripened ovary with attached and subtending reproductive structures.

Full-sib: Term used in population improvement. A full-sib family comprises progeny from a cross between two selected plants within the population.

Fumigation: Use of gas or vapours that sterilise soils or containers.

Fungicide: A chemical used to kill fungus.

Generation: One complete life cycle. The generation begins with the formation of the zygote and end when the resulting plant dies.

Genes: Basic units of hereditary material that dictate the characteristics of individuals.

Genetic sterility: A type of male sterility conditioned by nuclear genes. In contrast to cytoplasmic sterility, it may be transmitted by either the male or female parent.

Genotype: The true genetic makeup of a plant.

Genus: A group of species possessing fundamental traits in common but differing in other lesser characteristics.

Germination: The initiation of active growth by the embryo, resulting in the rupture of seed coverings and emergence of a new seedling plant capable of independent existence.

Gibberellin: A group of compounds that naturally controls stem elongation.

Grafting: The process of joining a stem or bud of one plant on to the root stock of a different plant so that by tissue regeneration they form a union and grow as one plant.

Green manures: Green manuring are plant cover crops that are tilled into the soil.

Growing media: A material used to culture plants.

Growing season: The number of days between the average date of the last killing frost in spring and the first killing frost in fall.

Grow-out test: Test performed to determine the genuineness of seed as to species or variety.

Growth regulator: A chemical used to increase growth or shorter stems.

Gynodioecy: Plant species in which individuals are either female (pistillate flowers) or hermaphrodite (bisexual flowers). It is often regarded as an evolutionarily intermediate stage between bisexuality and dioecy.

Gynomonoecious: Refers to a plant species in which female and bisexual flowers are produced on the same plant.

Habit: Growth form or overall plant shape.

Habitat: Type of region in which a plant is native.

Hardening off: The process of gradually acclimatising tender plants, greenhouse or indoor grown plants to outdoor growing conditions or survive a more adverse environment.

Hardiness/hardy: The ability of a plant to withstand low temperatures or frost, without artificial protection.

Hardpan: A layer in the soil that cannot be penetrated and which restricts root penetration as well as movement of air and water.

Heading back: Cutting an older branch or stem back to a stub or twig.

Healing: The process of closing of a plant wound or graft union.

Hectare: Standard area measurement in the metric system and it equals to 10,000 square meters in area.

Herbicide: A material or chemical kills essentially all plants or be quite selective in their activity.

Hermaphrodite: Papaya plants bear perfect flowers and correctly termed as andromonoecy, which indicates the occurrence of staminate and hermaphroditic flowers on the same plant.

Heterozygous: Having unlike alleles at corresponding loci of homologous chromosomes. An organism may be heterozygous for one or several genes.

High-pressure sprayer: A sprayer using a high-pressure pump to force the spray through nozzles for both creating a fine spray and delivery to the plant.

Homozygous: Having genes at corresponding loci on homologous chromosomes that are identical.

Honey dew: The sticky secretion produced by sucking insects such as aphids.

Humus: Organic matter rich in nutrients, brown or black in colour resulting from partial decay of leaves and other matter.

Hybrid: The offspring of two plants of different species or varieties of plants. Hybrids are created when the pollen from one kind of plant is used to pollinate an entirely different variety, resulting in a new plant altogether.

IAA: Indole-3-acetic acid; used to promote rooting in cuttings.

Imperfect flower: A flower that lacks either stamens or pistils.

Inbred/inbreeding: An undesirable plant made by the crossing of two related plants.

Incomplete flower: Flower that lacks any one or more of the floral parts, namely, calyx, corolla, stamens and pistils.

Indeterminate: Said of those kinds of inflorescence whose terminal flowers open last; hence the growth or elongation of the main axis is not arrested by the opening of the first flowers.

Indicator plant: A sensitive plant that alerts to some conditions such as virus, drought, and so on.

Infection: At the epidemiological level of the pathosystem, infection refers to contact made between the host and the parasite known as auto-infection and allo-infection. At the histological level, infection refers to the process of penetration of a host by a pathogen.

Inorganic: Something that was never alive and manufactured by a human.

Inter-cropping: When two crops of different height, canopy, adaptation, root system and growth habit are made to grow simultaneously in such a way that they accommodate each other with least competition.

Inter-specific: Between species or hybrid cross.

Inter-stem/inter-stock: A piece of stem tissue grafted between a rootstock and a scion.

Introduced: A plant that is exotic or brought from another region.

Isolation: The act of keeping the seed crops away from the sources of physical and genetical contamination.

Isolation distance: The distance to be maintained between the seed crop and the contaminant.

Juvenile: An early non-reproductive phase of plant growth, usually characterised by different leaf shapes or by non-flowering and vigorous increase in size, and often thorniness.

Larva: Immature insects such as caterpillars.

Layering: A method of vegetative propagation by which a branch of a plant is rooted while still attached to the plant by securing it to the soil with a piece of wire or other means.

Leaching: The downward movement of nutrients or salts through the soil profile in soil water. Leaching accounts for nutrient losses but can also be beneficial in ridding a soil of excess salts.

Leaf spot: One of many types of fungal or bacterial diseases causing round marks.

Loam: A rich soil composed of clay, sand and organic matter.

Male sterile: Describes the complete or partial failure of a male plant to produce mature reproductive pollen cells.

Manure: Organic matter that is excreted by animals and used as a soil amendment and fertiliser.

Maturation: The stage of development leading to the attainment of physiological or horticultural maturity.

Mature: A later phase of plant growth characterised by flowering, fruiting and reduced rate of size increase.

Media/soil media: Any substrate used to hold plant roots.

Microclimate: Variations of the climate within a given area and a very local or small-scale climate such as valley or woodland.

Micronutrient: An essential nutrient that is needed in small amounts and also called a trace or minor element.

Mixed cropping: Two or more than two crops are grown in mixed stand either as mixture or in separate rows in different proportion. When any of the crops in mixed stand gets vitiated due to one reason or other, the other crop may act as an insurance against complete loss or other.

Monoculture: Growing only one particular type of plant.

Monoecious: A species with unisexual flowers, having both male and female sexes on the same plant.

Morphology: The study of structure or forms.

Mother block: A certified group of plants used to derive propagation tissues or scion.

Moulds: Multicellular organisms that form fuzzy or powdery patches (mycelium) on organic matter such as fruits and vegetables. A type of fungus that grows on decaying plant tissue.

Mulch: A material applied to the surface of a soil for a variety of purposes such as conservation of moisture, stabilisation of soil temperature, suppression of weed growth, and so on.

Mutation: A sudden change in the expected genetic or tissue makeup of a plant resulting in an altered individual, generally disadvantageous to the mutated plant's survival.

Native plant: Native means original to an area. Any plant that occurs and grows naturally in a specific region or locality.

Naturalised: Thoroughly established but originally from a foreign area.

Nematode: Thread-like round worms that live in soil and water but also live in plants.

Niacin: Type of vitamins that dilates blood vessels, thus increasing circulation and reducing high blood pressure. It significantly lowers blood cholesterol and triglycerides. It improves resistance to stress, regulates blood sugar, treats dizziness and ringing in the ears.

Node: The point where a leaf or other structure meets a stem node or part of a stem from which a leaf or new branch starts to grow.

Nodule: Round bacteria-filled swellings on the roots of legume plants.

Nomenclature: The study of naming plants including spelling and format.

NPK: Nitrogen, phosphorous and potash are three main food elements necessary for plant life. Nitrogen is required for foliage and growth, phosphorous for flower and potash for roots.

Nutrient: A chemical required by a plant for growth.

Objectionable weed: Weeds whose seeds are difficult to be separated once mixed with crop seed or which are poisonous or injurious or are having a smothering effect on the main crop or are difficult to eradicate once established or are having a high multiplication ratio thus making their spread quick.

Off type (rogue): Plant that does not conform to the varietal characteristics described by the breeder. To designate a plant as an off type/rogue it is not necessary to identify it as to variety.

Open pollinated: A plant that is pollinated by natural means, such as bees and other insects, animals or the wind.

Organic matter: Materials rich in carbon of either plant or animal origin, which exist in all stages of decomposition of soils.

Outcross: The mating of a hybrid with a third parent and produce an off-type plant resulting from pollen of a different sort contaminating a seed field.

Ovary: The ovule-bearing part of a pistil. The female part of a flower containing immature seeds (ovules).

Ovule: The egg-containing unit of an ovary, which becomes the seed after fertilisation.

Pantothenic acid: It is considered as the 'Anti-Stress' vitamin because of its important role in the functioning of the adrenal glands, which produce hormones that help our bodies respond to stress. It plays an important role in making haemoglobin, which transports oxygen throughout our bodies. It is also helpful in detoxifying harmful chemicals such as herbicides and insecticides.

Papain: An enzyme extracted from the papaya fruit that aids the digestion of protein by breaking them down into smaller peptones. It has also been used to prevent several disorders and diseases.

Pathogen: An organism that causes a plant disease.

Peatmoss: An organic soil additive from Sphagnum and related mosses. This is a good water retentive addition to the soil but tends to add the acidity of the soil pH.

Pedicel: The stalk of a flower or fruit when in a cluster or solitary.

Peduncle: The stalk of a flower cluster or a single flower when the flower is solitary, or the remaining member of a reduced inflorescence.

Pentadria type flower: In severe cases, five antepetalous stamens are completely transformed into carpels and the resulting flower resembles a female one with a rounded ovary and free petals almost all along their length.

Perfect flower: A flower having both functional stamens and pistils or a plant with both functioning male and female parts.

Pericarp: The fruit wall consisting of three distinct layers, namely, the exocarp, the mesocarp and the endocarp.

Perlite: A light weight white-coloured soil additive from volcanic materials.

Pest: Any species, strain or biotype of plant, animal or pathogenic agent injurious to plants or plant products.

Pest management: The control of a pest or group of pests by a broad range of techniques from biological control to pesticides. The goal is to keep damage below economic levels without completely eliminating the pest.

Petals: Structures collectively making the corolla, which protect the inner reproductive structures and often attract insects by either their colour or their nectar and thus facilitate pollination.

pH: The scale used to denote the acidity or alkalinity (lime content) of a soil. A pH of 7 is neutral, less than 7 is acidic, more than 7 is basic and 6.5 supports most plant life.

Phenotype: The visual appearance of a plant regardless of genetics.

Physiological maturity: The stage of development when a plant or plant part will continue ontogeny even if detached.

Pistil: The female reproductive organ consists of stigma, style and ovary.

Pistillate: An imperfect flower with a pistil, or seed organ, but having no functional stamens.

Plant growth: A permanent increase in volume, dry weight or both.

Plant pathogen: A microorganism that causes a plant disease.

Plant propagation: Increase in numbers or perpetuation of a species by reproduction.

Planting ratio: The recommended ratio in which the male and female plants are to be planted to make a crossing block in hybrid seed production.

Pollen: The male cells or microspores produced by the stamens.

Pollen parent: Parent that furnishes the pollen that fertilises the ovule of the other parent in the production of seed.

Pollination: The transfer of pollen from the stamen to the pistil, which results in formation of a seed.

Pollinator: Line or population used as a male parent or pollen donor.

Polycarpic: Flowering and fruiting takes place many times.

Polygamous: Bearing unisexual and bisexual flowers on the same plant.

Polyploid: Plant having other than the diploid (2n) number of chromosomes.

Precocity: Tendency of a given species/cultivar to mature early.

Producer: A person who grows or distributes certified seed in accordance with the procedures and standards of the certification agency.

Protogynous: Having the stigma receptive to pollen before the pollen is released from the anthers of the same flower.

Pure seed: Seed of all botanical varieties of each species under analysis.

Purity: The relative stability and uniformity of a breeding line.

Recessive gene: A gene not expressed in the heterozygous state when a dominant gene is present at the same locus on the other homologous chromosome or a gene masked by the effect of another specifically by a dominant allele.

Recurrent: A plant more or less blooming in two or more sessions or months.

Registered: A cultivar catalogued with an International Registrar (IRA).

Relative humidity: The measurement of the amount of moisture in the atmosphere.

Ripening: The composite of the processes that occur from the latter stages of growth and development through the early stages of senescence and that results in characteristic aesthetic and/or food quality, as evidenced by changes in composition, colour, texture, or other sensory attributes.

Root: Vegetative plant part that anchors the plant, absorbs water and minerals in solution, and often stores food.

Rooting hormone: A growth hormone in powder or liquid form, which promotes the formation of roots at the base of a cutting.

Scion: Desirable tissue or a short length of stem, taken from one plant which is then grafted onto the rootstock of another plant.

Scorch: Injury to plant parts due to burning by hot wind or water loss. Sometimes it is also used as a synonym of scald and sun burning.

Seed: A fertilised ripened ovule that contains an embryo or a ripened plant ovary capable of germinating to produce another plant. Plant embryo with associated stored food encased in a protective seed coat.

Seed borne disease: Disease that is carried either within the seed, i.e. internally seed borne or on the seed, i.e. externally seed borne or both.

Seedling: A young plant grown from a seed.

Self: To cross a plant with its own flowers or one of identical type; said of flowers of one colour as opposed to a bicolour.

Self-fertile: A plant capable of producing viable seeds with its own pollen.

Self-fertilisation: Union of an egg with a sperm from the same flower or from another flower on the same plant or within a clone.

Self-pollination: Transfer of pollen from an anther to the stigma of the same flower or another flower on the same plant or within a clone.

Senescence: That process follows physiological maturity or horticultural maturity and lead to death of tissue.

Sepal: Structures that usually form the outermost whorl of the flower, collectively, the calyx.

Sib: Short hand for sibling or sister/brother in a seedling population.

Sibbing: Cross between plants from the same population. Generally, pollen is collected from several plants from the same population, bulked and crossed onto sister plants of the same population.

Slow-release fertiliser: A fertiliser that is made by coating the particles with a wax or other insoluble or very slowly soluble material to provide a predictable, slow release of the encapsulated materials.

Soil: The outer, weather layer of the earth's crust that has the potential to support plant life. Soil is made up of inorganic particles, organic matter, microorganisms, water and air.

Soil drench: A media treatment to kill fungi.

Soil management: The practices used in treating a soil, which may include various types of tillage and production systems.

Species: A natural group of plants composed of similar individuals that can produce similar offspring, usually including several minor variations. A unit of botanical classification capable of reproducing itself.

Stamen: The male part of a flower composed of the anther and the filament.

Staminate: An imperfect flower with stamens or pollen-producing structures, but with no pistil, or seed-producing structure.

Stigma: The pollen-receiving site of the pistil.

Strain: A group of similar individuals within a variety. Advanced generation random-mating population derived from a few selected inbred lines. A variety produced by crossing inter-se a number of inbred lines (usually five to eight) selected for their good general combining ability. The variety is subsequently maintained by open pollination.

Succulent: A fleshy plant that hold water in itself, usually in the leaves or stem. Thickened, juicy, fleshy tissues that is more or less soft in texture.

Systemic pesticide: A chemical that is absorbed by a plant and is translocated in its vessels, either to kill feeding insects on the plant, or to kill the plant itself.

Test cross: Cross made with a homozygous recessive parent to determine whether an individual is homozygous or heterozygous.

Tetraploid: Polyploidy plant having four sets of identical chromosomes (4n).

Thinning: Removing excess seedlings, to allow sufficient room for growth and better size and quality in the remaining plants. Removing stems, branches and fruits to give the plant a more open structure.

Tissue culture: The growing of masses of unorganised cells on agar or in liquid suspension. Useful for the rapid asexual multiplication of plants.

Top-dressing: A fertiliser or compost applied at the soil level.

Transplant: A seedling that grows its first true leaves, which resemble the adult plant leaves rather than the seed leaves that first appear.

Transplanting: The process of digging up a plant and moving it to another location.

Unisexual flowers: Flowers having either pistils or anthers, but not both the organs.

Variant: A variation between strains. A plant that is genetically different from the wild plant.

Variety: A sub-division of a kind identifiable by growth, yield, plant, fruit, seed or other characteristics. It also denotes on assemblage of cultivated individuals, which are distinguished by a character (morphological, cytological, chemical or others) significant for the purposes of agriculture, or horticulture and which when reproduced (sexually or asexually) or reconstituted retain their distinguishing features.

Vermiculite: The mineral mica, heated to the point of expansion. A good addition to container potting mixes, vermiculite retains moisture and air within the soil.

Virus: A deforming microorganism that causes disease and death.

Vitamin A: ß-carotene is a safe non-toxic form of vitamin. It is a powerful antioxidant that helps protect the cells against cancer by neutralising free radicals, necessary for new cell growth. It counteracts night blindness and weak eyesight and builds resistance to infections. It slows the progression of osteoarthritis and cataracts, and helps prevent macular degeneration of the eyes.

Vitamin B_1: Thiamin is known as the 'Morale Vitamin' because of its beneficial effects on the nervous system and mental attitude. It enhances circulation, assists in blood formation, carbohydrate metabolism and digestion. It plays a key role in generating energy and promotes good muscle tone. It acts as an antioxidant, protecting the body from degenerative effects of aging.

Vitamin B_{12}: It is also known as cynocobalamin, which helps in the formation of red blood cells, thus helping prevent anaemia. It increases energy levels promoting a healthy immune system and nerve function. It is required for the proper digestion of foods, the synthesis of protein and carbohydrates and fat metabolism. It aids in cell formation and cellular longevity.

Vitamin B-2: Commonly known as riboflavin, it is necessary for red blood cell formation, antibody production, cell respiration and growth. It alleviates eye fatigue and is important in the prevention of cataracts. It aids in the metabolism of carbohydrates, fats and proteins. It promotes the oxygenation of the skin, hair and nails. It has also been used to treat dandruff. It aids in the release of energy from food and reduces the occurrence of migraine headaches. It helps eliminate cracked mouth, lips, tongue and supports the production of adrenal hormones.

Vitamin B_6: Pyridoxine can significantly reduce the risk of heart disease by inhibiting the formation of homocysteine. It is a toxic chemical that attacks the heart muscle and allows the deposition of cholesterol around the heart muscle. It aids in maintaining the central nervous system and normal brain function. It reduces muscle spasms, leg cramps and stiffness of the hands. It relieves nausea and migraines, lowers cholesterol, improves vision and aids in the prevention of PMS.

Vitamin C: Ascorbic acid is a major and very potent antioxidant. It plays a primary role in the formation of collagen, which is important for the growth and repair of body tissue, cells, gums, blood vessels, bones and teeth. It protects against the harmful effects of pollution, infection and enhances the immune system.

Vitamin D: Its deficiency plays a key role in the development of type 1 diabetes. It is needed for islet cells to produce insulin. The hormone that allows cells to take up blood sugar. Without enough Vitamin D, islet cells do not produce insulin. It helps regulate white blood cells that make up the immune system.

Vitamin E: Tocopherol is a 'super' antioxidant, which protects cells against damage caused by 'free radicals'. It is extremely important in the prevention of cancer and cardiovascular disease. It is also useful in treating premenstrual syndrome and fibrocystic disease of the breast.

Viviparous: Germinating or sprouting from seed or bud while still attached to the parent plant.

Vivipary/viviparous: With live plantlets on a mother plant.

Volunteer plant: A plant of the crop species same as that of the seed crop, which comes up on account of self-seeding from the previous season's crop.

Waterlogged: Soil that is oversaturated with water.

Weed: A plant growing where it is not wanted.

Wild types: Naturally occurring non-domesticated crop relatives.

X: Indicates a hybrid.

Zinc: Zinc is needed to make important antioxidant enzymes and is essential for protein synthesis and collagen formation. It governs the contractibility of muscles, helps in the formation of insulin and helps prevent macular degeneration (one of the most common causes of vision loss in the elderly) and helps prevent the onset of cataracts. It is important for blood stability exerts a normalising effect on the prostate and development of all reproductive organs. Zinc is critical in the male sex drive and is involved in hormone metabolism, sperm formation and sperm motility. It helps prevent and reduce the length and severity of the common cold.

Zone: An area distinguished by a range of annual average minimum temperatures and used in describing plant hardiness. Regions that share similar climatic and rainfall conditions producing similar growing seasons.

Abbreviations

⚥	Hermaphrodite
♀	Pistillate
♂	Staminate
AFLP	Amplified fragment length polymorphism
BAC	Bacterial artificial chromosome
B/C ratio	Benefit–cost ratio
BC_6	Sixth generation of backcrossing
BG	Benzyl glucosinolate
BITC	Benzyl isothiocyanate
cM	Centi Morgan
CO	Coimbatore
cp	Coat protein
CPCT	Center for Protected Cultivation Technology
DAF	DNA amplification fingerprinting
DAPI	4, 6-Diamidino-2-phenylindole
DF	Dengue fever
DRDO	Defence Research and Development Organization
ELISA	Enzyme linked immunosorbent assay
F	Female
Fig	Figure
FISH	Fluorescence *in-situ* hybridisation
GCA	General combining ability
GE	Genetically Engineered
GnHCl	Guanidium hydrochloride
H	Hermaphrodite
HDL-C	High density lipoprotein cholesterol
IAA	Indole acetic acid
IBA	Indole-3-butyric acid
ISSR	Intersimple sequence repeat amplified
KNO_3	Potassium nitrate
MAS	Marker-assisted selection
MH	Maleic hydrazide
MS	Murashige and Skoog
NEPZ	North Eastern Plains Zone
PLT	Platelet count
PPFD	Photosynthetic photon flux density
PRSV	Papaya ringspot virus
RAPD	Randomly amplified polymorphic DNA
S	Staminate

SCA	Specific combining ability
SSR	Simple sequence repeat markers
SuF	Suppressing femaleness
TDZ	Thiodiazuron
TSS	Total soluble solids

References

Abdulazeez, A.M., D. A. Ameh, S. Ibrahim, J. Ayo and S. F. Ambali. 2009. Effect of fermented and unfermented seed extracts of *Carica papaya* on pre-implantation embryo development in female Wistar rats (*Rattus norvegicus*). *Scientific Research Essay* 4(10):1080–84.

Abdulazeez, M.A. 2008. Effect of fermented and unfermented seed extract of *Carica papaya* on implantation in Wistar rats (*Rattus norvegicus*). Thesis submitted to Department of Biochemistry, A.B.U Zaria.

Abdullah, M., P. S. Chai, C. Y. Loh et al. 2011. *Carica papaya* increases regulatory T cells and reduces IFN-γ+ CD4+ T cells in healthy human subjects. *Molecular Nutritional and Food Research* 55(5): 803–06.

Abu-Alruz, K., A. S. Mazahreh, J. M. Quasem, R. K. Hejazin and J. M. El-Qudah. 2009. Effect of proteases on meltability and stretchability of nabulsi cheese. *American Journal of Agriculture and Biological Sciences* 4: 173–78.

Adebiyi, A., P. G. Adaikan and R. N. V. Prasad. 2003. Tocolytic and toxic activity of papaya seed extract on isolated rat uterus. *Life Sciences* 74: 581–92.

Adetuyi, F. O., L. T. Akinadewo, S. V. Omosuli and A. Lola. 2008. Antinutrient and antioxidant quality of waxed and unwaxed pawpaw (*Carica papaya*) fruit stored at different temperatures. *African Journal of Biotechnology* 7: 2920–24.

Ahamefula Sunday, E., E. Ify and K. O. Uzoma. 2014. Hypoglycemic, hypolipidemic and body weight effects of unripe pulp of *Carica papaya* using diabetic Albino rat model. *Journal of Pharmacognosy and Phytochemistry* 2(6): 109–14.

Ahmad, N. and M. Anis. 2007. Rapid plant regeneration protocol for cluster bean (*Cyamopsis tetragonoloba* L.Taub.). *Journal of Horticulture Sciences and Biotechnology* 84: 585–89.

Ahmad, I., S. Bhagat, T. V. R. S. Sharma, K. Kumar, P. Simachalam and R. C. Srivastava. 2010. ISSR and RAPD marker based DNA fingerprinting and diversity assessment of *Annona* spp. in South Andaman. *Indian Journal of Horticulture* 67: 147–51.

Ahmad, M. A. and K. Sultana. 1981. Studies on chemical control of root-knot nematodes. In *Proceeding of 3rd Research Planning Conference on Root-Knot Nematodes*, Meloidogyne spp. Jakarta, Indonesia, Raleigh, NC, USA: North Carolina State University, pp. 130–131.

Ahmad, N., H. Fazal, M. Ayaz, B. H. Abbasi, I. Mohammad and L. Fazal. 2011. Dengue fever treatment with *Carica papaya* leaves extracts. *Asian Pacific Journal of Tropical Biomedicines* 1: 330–33.

Ahmed, J., U. S. Shivhare and K. S. Sandu. 2002. Thermal degradation kinetics of carotenoids and visual color of papaya puree. *Journal of Food Sciences*, 67: 2692–95.

Akinboro, A. and A. A. Bakare. 2007. Cytotoxic and genotoxic effects of aqueous extracts of five medicinal plants on *Allium cepa* Linn. *Journal of Ethnopharmacology* 112(3): 470–75.

Akinyemi, S. O. S., J. O. Makinde, I. O. O. Aiyelaagbe, F. M. Tairu and O. O. Falohun. 2004. Growth and yield response of 'Sunrise Solo' papaya to weed management Strategies. New directions for a diverse planet. *Proceeding of the 4th International Crop Congress*, Brisbane, Australia.

Allan, P. 2005. Phenology and production of *Carica papaya* 'Honey Gold' under cool subtropical conditions. *Acta Horticulturae* 740: 217–23.

Allan, P. and C. N. MacMillan. 1991. Advances in propagation of *Carica papaya* L. cv. Honey Gold cuttings. *Journal of the South African Society for Horticultural Science* 1: 69–72.

Allan, P., J. McChlery and D. Biggs. 1987. Environmental effects on clonal female and male *Carica papaya* L. plants. *Scientia Horticulturae* 32: 221–32.

Alobo, A. P. 2003. Proximate composition and selected functional properties of defatted papaya (*Carica papaya* L.) kernel flour. *Plant Foods for Human Nutrition* 58: 1–7.

Amri, E. and F. Mamboya. 2012. Papain, a plant enzyme of biological importance: A review. *American Journal of Biochemistry and Biotechnology* 8(2): 99–104.

Anandan, R, S. Thirugnanakumar, D. Sudhakar and P. Balasubramanian. 2011. *In vitro* organogenesis and plantlet regeneration of (*Carica papaya* L.). *Journal of Agriculture Technology* 7(5): 1339–48.

Anwar, M., D. D. Patra, S. Chand, K. Alpesh, A. A. Naqvi and S. P. S. Khanuja. 2005. Effect of organic manures and inorganic fertilizer on growth, herb and oil yield, nutrient accumulation, and oil quality of *French basil. Communications in Soil Science and Plant Analysis* 36(13–14): 1737–46.

APS. 2014. Diseases of papaya (*Carica papaya* L.) The American Phytopathological Society. Available at http://en.wikipedia.org/wiki/List_of_papaya_diseases.

Aravind, G., D. Bhowmik, S. Duraivel and G. Harish. 2013.Traditional and medicinal uses of *Carica papaya*. *Journal of Medicinal Plants Studies* 1(1): 7–15.

Arkle Junior, T. D. and H. Y. Nakasone. 1984. Floral differentiation in the hermaphroditic papaya. *HortScience* 19: 832–34.

Aryantha, I. P., R. Cross and D. I. Guest. 2000. Suppression of *Phytophthora cinnamomi* in potting mixes amended with uncomposted and composted animal manures. *Phytopathology* 90: 775–82.

Asudi, G. O., F. K. Ombwara, F. K. Rimberia, A. B. Nyende, E. M. Ateka and L. S. Wamocho. 2013. Evaluating diversity among Kenyan papaya germplasm using simple sequence repeat markers. *African Journal of Food, Agriculture Nutrition and Development* 13(1): 7307–24.

Awada, M. 1958. Relationships of minimum temperature and growth rate with sex expression of papaya plants (*Carica papaya* L.). *Hawaii Agricultural Experiment Station, Technical Bulletin* 38: 1–16.

Awada, M. and W. Ikeda. 1957. Effects of water and nitrogen application on composition, growth, sugars in fruits, yield, and sex expression of the papaya plants (*Carica papaya* L.). *Hawaii Agricultural Experimental Station, Technical Bulletin* No. 33.

Azad, M. A. K. and M. G. Rabbani. 2004. Studies on floral biology of different *Carica* species. *Pakistan Journal of Biological Sciences* 7: 301–04.

Azarkan, M., A. Amrani, M. Nijs et al. 1997. *Carica papaya* latex is a rich source of a class II chitinase. *Phytochemistry* 46(8): 1319–25.

Azarkan, M., A. El Moussaoui, D. van Wuytswinkel, G. Dehon and Y. Looze. 2003. Fractionation and purification of the enzymes stored in latex of *Carica papaya*. *Journal of Chromatography B Analytical Technology, Biomedical and Life Sciences* 790(1–2): 229–38.

Azene, M., T. S. Workneh and K. Woldetsadik. 2011. Effects of different packaging materials and storage environment on postharvest quality of papaya fruit. *Journal of Food Science and Technology* 51(6): 1041–55.

Babu, A. R., D. S. Rao and M. Parthasarathy. 2003. *In sacco* dry matter and protein degradability of papaya (*Carica papaya*) pomace in buffaloes. *Buffalo Bulletin* 22: 12–15.

Badillo, V. M. 1971. *Monografia de la familia* Caricacae, Editorial Nuestra, America C.A., Maracay, Venezuela, pp. 111.

Badillo, V. M. 2001. Nota correctiva *Vasconcella* St. Hilly no *Vasconcella* (Caricaceae). *Ernstia* 11: 75–76.

Balakrishnan, K., T. N. Balamohan, L. Veerannah, M. Kulasekaran and K. G. Shanmughavelu. 1986. Seed development and maturation in papaya. *Progressive Horticulture* 18: 68–70.

Bamisaye, F. A., E. O. Ajani and J. B. Minari. 2013. Prospects of ethnobotanical uses of pawpaw (*Carica papaya*). *Journal of Medicinal Plants Studies* 1(4): 171–77.

Bankapur, V. M. and A. F. Habib. 1979. Mutation studies in papaya *(Carica papaya* L.)-ratio sensitivity. *Mysore Agricultural Sciences* 13: 18–21.

Bari, L., P. Hassen, N. Absar et al. 2006. Nutritional analysis of two local varieties of papaya (*Carica papaya*) at different maturation stages. *Pakistan Journal of Biological Sciences* 9: 137–40.

Bau, H. J., Y. J. Kung, J. A. J. Raja et al. 2008. Potential threat of a new pathotype of papaya leaf distortion mosaic virus infecting transgenic papaya resistant to papaya ringspot virus. *Phytopathology* 98(7): 848–56.

Baxy, J. N. 2009. *Morphological characterization of four locally available papaya cultivars.* B.Sc. (Hons) Dissertation, UOM Library, University of Mauritius.

Beeley, J. A., H. K. Yip and A. G. Stevenson. 2000. Chemo chemical caries removal: A review of the techniques and latest developments. *British Dental Journal – Nature* 188: 427–30.

Benchimol, L. L., C. L. Souza Jr. and A. P. Souza. 2005. Microsatellite-assisted backcross selection in maize. *Genetics and Molecular Biology* 28: 789–97.

Benson, C. W. and M. Poffley. 1998. Growing pawpaws. Agnote 386 no. D8. Northern Territory Government, Department of Primary Industry, Fishery and Mines.

Bertan, I., F. I. F. Carvalho, A. C. Oliveira and G. Benin. 2009. Morphological, pedigree, and molecular distances and their association with hybrid wheat performance. *Pesquisa Agropecuária Brasileira* 44: 155–62.

Bhattacharya, J. and S. S. Khuspe. 2001. *In vitro* and *in vivo* germination of papaya (*Carica papaya* L.) seeds. *Scientia Horticulturae* 91: 39–49.

Bhattacharya, J., S. S. Khuspe, N. N. Renukdas and S. K. Rawal. 2003. Somatic embryogenesis and plant regeneration from immature embryo explants of papaya cv. Washington and Honey Dew. *Indian Journal of Experimental Biology* 40: 624–27.

Biswas, B. C. 2010. Success stories of papaya farmers. *Fertilizer Marketing News* 41(2): 15–18.

Blanco, C., N. Ortega, C. Rodolfo, M. Alvarez, A. G. Dumpierrez and T. Carillo. 1998. *Carica papaya* pollen allergy. *Annals of Allergy Asthma Immunology* 81: 171–75.

Blazevic, L., A. Radonic, J. Mastelic, M. Skocibusic and A. Maravic. 2010. Glucosinolates, glycosidically bound volatiles and antimicrobial activity of *Aurinia sinuate (Brassicaceae). Food Chemistry* 121: 1020–28.

Bose, T. K. and S. K. Mitra. 1990. *Fruits: Tropical and Subtropical.* Calcutta, India: Naya Prakash.

Boshra, V. and A.Y. Tajul. 2013. Papaya—An innovative raw material for food and pharmaceutical processing industry. *Health Environment Journal* 4(1): 68–75.

Brekke, J. E., C. G. Cavaletto, T. O. M. Nakayama and R. Suehina. 1976. Effects of storage temperature and container lining on some quality attributes of papaya nectar. *Journal of Agricultural and Food Chemistry* 24: 341–43.

Brekke, J. E., H. T. Chan Jr. and C. G. Cavaletto. 1972. Papaya puree: A tropical flavor ingredient. *Food Production and Development* 6: 36–37.

Bron, H. V. and A. P. Jacomino. 2006. Ripening and quality of Golden papaya fruit harvested at different maturity stages. *Brazilian Journal of Plant Physiology* 18: 389–96.

Brown, J. E., J. M. Bauman, J. F. Lawrie, O. J. Rocha and R. C. Moore. 2012. The structure of morphological and genetic diversity in natural populations of *Carica papaya* (Caricaceae) in Costa Rica. *Biotropica* 44(2): 179–88.

Bugbee, B. and O. Monje. 1992. The limits of crop productivity: Theory and validation. *BioScience* 42(7): 494–502.

Buisson, D. and D. W. Lee. 1993. The developmental responses of papaya leaves to simulated canopy shade. *American Journal of Botany* 80(8): 947–52.

Caberera-Ponce, J. L., A. Vegas-Garcia and L. Herrera-Estrella. 1995. Herbicide resistant transgenic papaya plants produced by an efficient particle bombardment transformation method. *Plant Cell Reports* 15(1–2): 1–7.

Cabral, A. A., H. G. Pulido, B. R. Garay and A. G. Mora. 2008. Plant regeneration of *Carica papaya* L. through somatic embryogenesis in response to light quality, gelling agent and phloridzin. *Scientia Horticulturae* 118: 155–60.

Cai, W. Q., C. Gonsalves, P. Tennant et al. 1999. A protocol for efficient transformation and regeneration of *Carica papaya* L. *In vitro Cellular and Developmental Biology-Plant* 35(1): 61–69.

California Rare fruit Growers Inc. 1997. Papaya Carica Papaya L. California Rare fruit Growers Inc., Fullerton, California. Available at http://www.crfg.org/pubs/ff/papaya.html.

Campostrini, E. and D. M. Glenn. 2007. Ecophysiology of papaya: A review. *Brazilian Journal of Plant Physiology* 19: 413–24.

Cano, M. P., M. C. Lobo, B. D. Ancos and M. A. M. Galeazzi. 1996. Polyphenol oxidase from Spanish hermaphrodite and female papaya fruits (*Carica papaya* cv. Sunrise, Solo group). *Journal of Agriculture and Food Chemistry* 44(10): 3075–79.

Capoor, S. P. and P. M. Verma. 1961. Immunity to papaya mosaic virus in the genus *Carica*. *Indian Phytopathology* 14: 96–97.

Carneiro, C. E. and J. L. Cruz. 2009. Caracterizacao anatomica de orgaos vegetativos do mamoeiro. *Cienc Rural* 39(3): 918–21.

Carvalho, F. A. and S. A. Renner. 2013. The phylogeny of Caricaceae. In *Genetics and Genomics of Papaya,* eds. R. Ming and P. H. Moore. New York: Springer Science + Business Media.

Castilla, N. 1994. Greenhouses in the Mediterranean area: Technological level and strategic control of noxious animals and plants (OIBC/OILB), West Palaearctic regional management. *Acta Horticulturae* 361: 44–56.

Castillo, B., M. A. L. Smith and U. L. Yadava. 1997. Plant regeneration from encapsulated somatic embryos of *Carica papaya* L. *Plant Cell Reports* 17(3): 172–76.

Chadha, K. L. 1992. Scenario of papaya production and utilization in India. *Indian Journal of Horticulture* 49: 97–119.

Chakraborty, P., D. Ghosh, I. Chowdhury et al. 2005. Aerobiological and immunochemical studies on *Carica papaya* L. pollen: An aeroallergen from India. *Clinical Allergy* 60: 920–26.

Chan, Y. K., M. D. Hassan and U. K. Abu Bakar. 1999. Papaya: The industry and varietal improvement in Malaysia. In *The Papaya Biotechnology Network of Southeast Asia: Biosafety Considerations and Papaya Background Information*, eds. R. A. Hautea, Y. K. Chan, S. Attathom and A. F. Krattiger. Ithaca, New York: International Service for the Acquisition of Agri-biotech Applications.

Chan, Y. K. and P. Raveendranathan. 1984. Differential sensitivity of papaya varieties in expression of boron deficiency symptoms. *MARDI Research Bulletin* 12(3): 281–86.

Chand, S., M. Anwar and D. D. Patra. 2006. Influence of long-term application of organic and inorganic fertilizer to build up soil fertility and nutrient uptake in mint-mustard cropping sequence. *Communication of Soil Science and Plant Analysis* 37: 63–76.

Chandrika, U. G., E. R. Jansz, S. N. Wickramasinghe and N. D. Warnasuriya. 2003. Carotenoids in yellow- and red-fleshed papaya (*Carica papaya* L). *Journal of Science and Food Agriculture* 83: 1279–82.

Charoensiri, R., R. Kongkachuichai, S. Suknicom and P. Sungpuag. 2009. Beta-carotene, lycopene, and alpha-tocopherol contents of selected Thai fruits. *Food Chemistry* 113: 202–07.

Chaterlan, Y., G. Hernandez, T. Lopez et al. 2012. Estimation of the papaya crop coefficients for improving irrigation water management in south of Havana. *Acta Horticulturae* 928: 179–86.

Chavasit, V., R. Pisaphab, P. Sungpung, S. Jittinandana and E. Wasantwisut. 2002. Change in beta-carotene and vitamin-A contents of vitamin A-rich foods in Thailand during preservation and storage. *Journal of Food Sciences* 67: 375–79.

Chawla, P., A. Yadav and V. Chawla. 2014. Clinical implications and treatment of dengue. *Asian Pacific Journal of Tropical Medicines* 5(3): 169–78.

Chay-Prove, P., P. Ross, P. O'Hare et al. 2000. Agrilink series: Your growing guide to better farming. Pawpaw information kit. Queensland horticulture institute and department of primary industries, Qld, Nambour, Queensland.

Chen, C., Q. Yu, S. Hou et al. 2007. Construction of a sequence-tagged high-density genetic map of papaya for comparative structural and evolutionary genomics in Brassicales. *Genetics* 177: 2481–91.

Chen, M. H., P. G. Wang and E. Maida. 1987. Somatic embryogenesis and plant regeneration in *Carica papaya* L. tissue culture derived from root explants. *Plant Cell Reports* 6(5): 348–51.

Chen, Y. K. 1992. Progress in breeding of F_1 papaya hybrids in Malaysia. *Acta Horticulture* 292: 41–49.

Cheng, L. L., J. R. Nechols, D. C. Margolies et al. 2009. Foraging on and consumption of two species of papaya pest mites, *Tetranychus kanzawai* and *Panonychus citri* (Acari: Tetranychidae), by *Mallada basalis* (Neuroptera: Chrysopidae). *Environmental Entomology* 38(3): 715–22.

Chin, H. F., Y. L. Hor and M. B. Lassim. 1984. Identification of recalcitrant seeds. *Seed Science and Technology* 12: 429–36.

Chinoy, N. J., T. Dilip and J. Harsha. 2006. Effect of *Carica papaya* seed extract on female rat ovaries and uteri. *Phytotherapy Research* 9(3): 169–75.

Chinoy, N. J., J. M. D'Souza and P. Padman. 1994. Effects of crude aqueous extract of *Carica papaya* seeds in male albino mice. *Reproduction and Toxicology* 8(1): 75–79.

Chukwuemeka, N. O. and A. B. Anthoni. 2010. Antifungal effects of pawpaw seed extracts and papain on post-harvest *Carica papaya* L. fruit rot. *African Journal of Agriculture Research* 5: 1531–35.

Chuman, T., P. J. Landolt, R. R. Heath and J. H. Tumlinson. 1987. Isolation, identification, and synthesis of male-produced sex pheromone of papaya fruit fly, *Toxotrypana curvicauda* Gerstaecker (Diptera: Tephritidae). *Journal of Chemical Ecology* 13(9): 1979–92.

Clarindo, W. R., C. C. Roberto de, S. A. Fernanda, S. A. Isabella de and C. O. Wagner. 2008. Recovering polyploid papaya *in vitro* regenerants as screened by flow cytometry. *Plant Cell, Tissue and Organ Culture* 92: 207–14.

Claudinei, A. and A. A. Khan. 1993. Improving papaya seedling emergence by matriconditioning and gibberellin treatment. *Horticulture Science* 28(7): 708–09.

Clemente, H. S. and T. E. Marler. 2001. Trade winds reduce growth and influence gas exchange patterns in papaya seedlings. *Annals of Botany* 88(3): 379–85.

Codex, 2005. Codex standard for papaya. Codex Standard Series No. 183-1993, Reviews 1-2001, Amsterdam. 1-2005.

Cohen, L. W., V. M. Coghlan and L. C. Dihel. 1986. Cloning and sequencing of papain-encoding cDNA. *Gene* 48: 219–27.

Collard, E. and S. Roy. 2010. Improved function of diabetic wound-site macrophages and accelerated wound closure in response to oral supplementation of a fermented papaya preparation. *Antioxidant and Redox Signaling* 13(5): 599–606.

Conover, R. A., R. E. Litz and S. F. Malo. 1986. Cariflora: A papaya ringspot virus tolerant papaya for South Florida and the Caribbean. *Horticulture Science* 21(4): 1072.

Copland, M. J. W. and A. G. Ibrahim. 1985. Biology of glasshouse scale insects and their parasitoids. In *Biological Pest Control: The Glasshouse Experience*, eds. N. W. Hussey and N. Scopes. Ithaca: Cornell University Press, 240.

Couey, H. M., A. M. Alvarez and M. G. Nelson. 1984. Comparison of hot water spray and immersion treatments for control of postharvest decay of papaya. *Plant Diseases* 68: 436–37.

Coveness, F. E. 1967. Nematology studies 1960–1965. End of tour report MANR, Ibadan/USAID, p 135.

Crane, J. H. 2005. Papaya growing in the Florida home landscape. The Institute of Food and Agricultural Sciences (IFAS), University of Florida. *Horticulture Science* 11: 1–8.

Crocker, T. E. 1994. Propagation of fruit crops. Circular 456-A, Florida Cooperative Extension Service, *Institute of Food and Agricultural Sciences*, University of Florida, pp 1–8.

Cunha, R. J. P. and H. P. Haag. 1980. Mineral nutrition of papaya. In Fruit development and nutrient removal by harvesting. Anasis-da-Escola-Superior-de-Agricultura. *Luiz-de-Qeuiroz* 37: 169–78.

D'Amato, F. 1977. Cytogenetics of differentiation in tissue and cell cultures. In *Applied and Fundamental Aspects, Plant Cell, Tissue and Organ Culture*, eds. J. Reinert and Y. P. S. Bajaj. Berlin: Springer–Verlag, 343–57.

da Silva, J. A. T., Z. Rashid, D. T. Nhut et al. 2007. Papaya (*Carica papaya* L.) biology and biotechnology. *Tree, Forestry Science and Biotechnology* 1(1): 47–73.

Dai, B. B. 1960. Experiment on utilization of F_1 papaya hybrids. *Nung-yeh Yenchin Agricultural Research Taipai* 8: 17–29.

DAIS. 2009. Cultivating papayas. Department of Agriculture, Forestry and Fisheries, Directorate Agricultural Information Services Private Bag X144, Pretoria, 0001 South Africa. Available at www.nda.agric.za/publications.

Dakare, M. 2004. Fermentation of *Carica papaya* seeds to be used as "daddawa". An M.Sc. Thesis; Department of Biochemistry, A.B.U Zaria.

Damasceno Junior, P. C., T. N. S. Pereira, F. F. Silva, A. P. Viana and M. G. Pereira. 2008. Comportamento floral de híbridos de mamoeiro (*Carica papaya* L.) avaliados no verao e primavera. *Ceres* 55: 310–16.

Delph, L. F., A. M. Arntz, C. Scotti-Saintagne and I. Scotti. 2010. The genomic architecture of sexual dimorphism in the dioecious plant *Silene latifolia*. *Evolution* 64: 2873–86.

Deputy, J. C., R. Ming, H. Ma et al. 2002. Molecular markers for sex determination in papaya (*Carica papaya* L.). *Theoretical and Applied Genetics* 106: 107–11.

De Los Santos, R. F., L. E. N. Becerra, V. R. Mosqueda, H. A. Vasquez and A. B. Vargas. 2000. Manual de produccion de papaya en el estado de Veracruz. INIFAP-CIRGOC. Campo Experimental Cotaxtla. Folleto Tecnico Num. 17. Primera reedicion. Veracruz, Mexico.

de Oliveira, E. J., A. dos Santos Silva, F. M. de Carvalho et al. 2010. Polymorphic micro-satellite marker set for *Carica papaya* L. and its use in molecular-assisted selection. *Euphytica* 173: 279–87.

Desai, U. T. and A. N. Wagh. 1995. Papaya. In *Hand Book of Fruit Science and Technology: Production, Composition, Storage, and Processing,* eds. D. J. Salunke and S. S. Kadam. New York: Marcel Dekker, 4–314.

Desser, L., E. Zavadova and I. Herbacek. 2011. Oral enzymes as additive cancer therapy. *International Journal of Immunotherapy* 17(4): 153–61.

Devi, G. and P. L. Saran. 2014. Papaya cultivation: Gujarat's new success story. *Agriculture Today* 17(3): 50–52.

Dhaliwal, G. S. and M. I. S Gill. 1991. Testing pollen viability, germination and stigma receptivity in some papaya cultivars. *Journal of Research for Punjab Agricultural University* 28: 206–10.

Dharmarathna, S. L. C. A., S. Wickramasinghe, R. N. Waduge, R. P. V. J. Rajapakse and S. A. M. Kularatne. 2013. Does *Carica papaya* leaf-extract increase the platelet count? An experimental study in a murine model. *Asian Pacific Journal of Tropical Biomedicines* 3(9): 720–24.

Dhinesh Babu, K., R. K. Patel, A. Singh, D. S. Yadav, L. C. De and B. C. Deka. 2010. Seed germination, seedling growth and vigour of papaya under North East Indian condition. *Acta Horticulturae* 851: 299–305.

Dietrich, R. E. 1965. Oral proteolytic enzymes in the treatment of athletic injuries: A double-blind study. *Pennsylvania Medical Journal* 68: 35–37.

Dinesh, M. R., C. P. A. Iyer and M. D. Subramanyam. 1991. Combining ability studies in *Carica papaya* L. with respect to yield and quality characters. *Gartenbauwissenschaft* 56(2): 81–83.

Dinesh, M. R. and I. S. Yadav. 1998. Surya: A promising papaya hybrid. *Indian Horticulture* 43(3): 21–33.

Dobson, H., J. Cooper, W. Manyangarirwa, J. Karuma and W. Chiimba. 2002. *Integrated Vegetable Pest Management*. UK: Natural Resources Institute, University of Greenwich.

Downer, A., J. A. Menge and E. Pond. 2001. Association of cellulytic enzyme activity in eucalyptus mulch with biological control of *Phytophthora cinnamomi*. *Phytopathology* 91: 847–55.

Drea, S., L. C. Hileman, G. de Martino and V. F. Irish. 2007. Functional analyses of genetic pathways controlling petal specification in poppy. *Development* 134: 4157–66.

Drenth, A. and D. I. Guest. 2004. *Diversity and Management of Phytophthora in Southeast Asia*. ACIAR Monograph.

Drew, R. A., C. M. O'Brien and P. M. Magdalita. 1998. Development of *Carica* interspecific hybrids. *Acta Horticulture* 461: 285–91.

Drew, R. A. and N. G. Smith. 1986. Growth of apical and lateral buds of papaya (*Carica papaya* L.) as affected by nutritional and hormonal factors. *Journal of Horticulture Science and Biotechnology* 61: 535–43.

Duarte, I. A., J. M. Ferreira and C. N. Nuss. 2003. Potencial discriminatorio de tres testadores em top crosses de milho. *Pesquisa Agropecuária Brasileira* 38: 365–72.

Dubey, K. S. and R. P. Singh. 2012. Diseases and management of crops under protected cultivation. *Proceedings of the 26th Training Programme; Centre of Advanced Faculty Training in Plant Pathology*, G.B. Pant University of Agriculture and Technology, Pantnagar (Uttrakhand) India.

Duke, J. A. 1992. *Handbook of Phytochemical Constituents of GRAS Herbs and other Economic Plants*. Ann Arbor, MI: CRC Press, pp. 136–37.

Dwivedi, A. K. 1998. Coheritability and predicted response to selection in papaya (*Carica papaya* L.). *South Indian Horticulture* 46(1–2): 1–4.

Dwivedi, A. K., P. K. Ghanta and S. K. Mitra. 1998. Correlation and regeneration study of fruit production and its components in papaya (*Carica papaya* L.). *Horticulture Journal* 11(2): 29–32.

Dwivedi, A. K. and K. K. Jha. 1999. Relationship of pulp weight and its components in papaya (*Carica papaya* L.). *Journal of Research Birsa Agricultural University* 11(I):77–78.

Edwin, F. and M. V. Jagannadham. 2000. Single disulphide bond reduced papain exists in a compact intermediate state. *Biochemistry and Biophysics Acta* 1479: 69–82.

Edwin, F., M. Sharma and M. V. Jagannadham. 2002. Single disulfide bond reduced papain exists in a compact intermediate state. *Biochemistry and Biophysics Research and Communication* 1479: 69–82.

Elder, R. J., W. N. B. Macleod, K. L. Bell, J. A. Tyas and R. L. Gillespie. 2000. Growth, yield and phenology of 2 hybrid papaya (*Carica papaya* L.) as influenced by method of water application. *Australian Journal of Experimental Agriculture* 40: 739–46.

Ellis, R. H. 1984. Revised table of seed storage characteristics. *Plant Genetic Resouruces Newsletters* 58: 16–33.

Ellis, R. H., T. D. Hong and E. H. Roberts. 1990. An intermediary category of seed behaviour of coffee? *Journal of Experimental Botany* 41: 1167–74.

Ellis, R. H., T. D. Hong and E. H. Roberts. 1991. Effect of storage temperature and moisture on the germination of papaya seeds. *Seed Science Research* 1: 69–72.

El Moussaoui, A., M. Nijs, C. Paul et al. 2001. Revisiting the enzymes stored in the laticifers of *Carica papaya* in the context of their possible participation in the plant defense mechanism. *Cellular, Molecular and Life Sciences* 58: 556–70.

Eloisa, M., Q. Reyes and R. E. Paull. 1994. Skin freckles on solo papaya fruit. *Scientia Horticulturae* 58(1–2): 31–39.

Elujoba, A. A. 2001. Traditional medicine practice experience. (Personal communication).

Encyclopedia of Life (EOL), 2015. Available at http://eol.org/pages/585682/maps.

Erwin, D. C. and O. K. Ribeiro. 1996. *Phytophthora Diseases Worldwide*. St. Paul, Minnesota, USA: APS Press, The American Phytopathological Society.

Escudero, J., A. Acosta, L. V. Ramirez, I. B. Caloni and G. R. Sifre. 1994. Yield in three papaya genotypes and their tolerance to papaya ringspot virus, Puerto Rico. *Journal of Agriculture University of Puerto Rico* 78(3–4): 111–21.

Eustice, M., Q. Yu, C. W. Lai et al. 2008. Development and application of microsatellite markers for genomic analysis of papaya. *Tree Genetics and Genomes* 4: 333–41.

Evans, E. A. and F. H. Ballen. 2012. An overview of global papaya production, trade and consumption. *The Food and Resource Economics Department, Florida Cooperative Extension Service*, Institute of Food and Agricultural Sciences, University of Florida, Gainesville, FL FE913, 1–7 pp. Available at http://edis.ifas.ufl.edu.

Evans, E. A., F. H. Ballen and J. H. Crane. 2012. Cost estimates of establishing and producing papaya (*Carica papaya*) in South Florida. *Food and Resource Economics Department, Florida Cooperative Extension Service, Institute of Food and Agricultural Sciences*, University of Florida, Gainesville, Florida. FE 918:1-5. Available at http://edis.ifas.ufl.edu/.

Ewel, J. J. 1986. Designing agricultural ecosystems for the humid tropics. *Annual Reviews of Ecology and Systematic* 17(1): 245–71.

Fahey, J. W., A. T. Zalcmann and P. Talalay. 2001. The chemical diversity and distribution of glucosinolates and isothiocyanates among plants. *Phytochemistry* 56: 2–21.

FAOSTAT. 2012a. Papaya. Food and agriculture organization of the United Nations. FAOSTAT. FAO. Org.

FAOSTAT. 2012b. Crop Production. Available at http://faostat.fao.org/site/567/default.aspx#ancor.

Fernando, J. A., M. Melo and M. K. M. Soares. 2001. Anatomy of somatic embryogenesis in *Carica papaya* L. *Brazilian Achieves of Biology and Technology* 44: 247–55.

Fisher, J. R. 1980. The vegetative and reproductive structures of papaya *(Carica papaya* L.). *Lyonia* 1: 191–208.

Fitch, M. M. M. and R. M. Manshardt. 1990. Somatic embryogenesis and plant regeneration from immature zygotic embryos of papaya (*Carica papaya* L.). *Plant Cell Reports* 9(6): 320–24.

Fitch, M. M. M., R. M. Manshardt, D. Gonsalves, J. L. Slightom and J. C. Sanford. 1992. Virus resistant papaya plants derived from tissues bombarded with the coat protein gene of papaya ringspot virus. *Biotechnology* 10(11): 1466–72.

Fitch, M., P. Moore and T. Leong. 1998. Progress in transgenic papaya (*Carica papaya*) research: Transformation for broader resistance among cultivars and micropropagating selected hybrid transgenic plants. *Acta Horticulturae* 461: 315–19.

Flindt, M. L. 1979. Allergy to alpha-amylase and papain. *Lancet* 1: 1407–08.

Fouzder, S. K., S. D. Chowdhury, M. A. R. Howlider and C. K. Podder. 1999. Use of dried papaya skin in the diet of growing pullets. *British Poultry Sciences* 40: 88–90.

Freye, H. B. 1988. Papain anaphylaxis: A case report. *Allergy Proceeding* 9: 571–74.

García-Solis, P., E. M. Yahia, V. Morales-Tlalpan and M. Diaz-Munoz. 2009. Screening of antiproliferative effect of aqueous extracts of plant foods consumed in México on the breast cancer cell line MCF-7. *International Journal of Food Science and Nutrition* 60(6): 32–46.

Garrett, A. 1995. *The pollination biology of pawpaw (Carica papaya L.) in central Queensland*. Ph.D. Thesis, Central Queensland University, Rockhampton.

Ghanta, P. K., R. S. Dhua and S. K. Mitra. 1992. Response of papaya to foliar spray of boron, manganese and copper. *Horticultural Journal* 5(1): 43–48.

Ghosh, S. 2005. Physicochemical and conformational studies of papain/sodium dodecyl sulfate system inaqueous medium. *Colloids and Surfaces A: Physicochemical and Engineering Aspects* 264: 6–16.

Gill, L. S. 1992. *Ethnomedical Uses of Plants in Nigeria*, Benin, Nigeria: Uniben Press.

Goka, K. 1998. Mode of inheritance of resistance to three new acaricides in the Kanzawa spider mite *Tetranychus kanzawai* Kishida (Acari: Tetranychidae). *Experimental and Applied Acarology* 22: 699–708.

Gomez, M., F. Lajolo and B. Cordenunsi. 2002. Evolution of soluble sugars during ripening of papaya fruit and its relation to sweet taste. *Journal of Food Sciences* 67: 442–47.

Gonsalves, C., W. Cai, P. Tennant and D. Gonsalves. 1998. Effective development of papaya ringspot virus resistant papaya with untranslatable coat protein gene using a modified microprojectile transformation method. *Acta Horticulturae* 461: 311–14.

Gonsalves, C., D. R. Lee and D. Gonsalves. 2007. The adoption of genetically modified papaya in Hawaii and its implications for developing countries. *Journal of Development Studies* 43: 177–91.

Gonsalves, D. 1998. Control of papaya ringspot virus in papaya: A case study. *Annual Review of Phytopathology* 36: 415–37.

Gonsalves, D., S. Tripathi, J. B. Carr and J. Y. Suzuki. 2010. Papaya ringspot virus. *The Plant Health Instructor* DOI: 10.1094/PHI-I-2010-1004-01.

Grabowska-Joachimiak, A. and A. Joachimiak. 2002. C-banded karyotypes of two Silene species with heteromorphic sex chromosomes. *Genome* 252: 243–52.

Green, E., A. Samie, C. L. Obi, P. O. Bessong and R. N. Ndip. 2010. Inhibitory properties of selected South African medicinal plants against *Mycobacterium tuberculosis*. *Journal of Ethnopharmacology* 130: 151–57.

Guadalupe, G. I. J. 1981. Papaya cultivation in the collima region. *Circular IAB (Maxico)* 60: 60.

Guest, D. I., K. G. Pegg and A. W. Whiley. 1995. Control of Phytophthora diseases of tree crops using trunk-injected phosphonates. *Horticultural Reviews* 17: 299–330.

Guo, Z., A. McGill, L. Yu, J. Li and J. Ramirez. 1996. ChemInform abstract: S-nitrosation of proteins by *N*-methyl-*N*-nitrosoanilines. *ChemInform* DOI: 10.1002/chin.199628113.

Guo, Z., J. Ramirez, J. Li and P. G. Wang. 1998. Peptidyl N-nitrosoanilines: A novel class of cysteine protease inactivators. *Journal of The American Chemical Society* 120: 3726–34.

Gupta, O. P., N. Sharma and D. Chand. 1992. A sensitive and relevant model for evaluating anti-inflammatory activity-papaya latex-induced rat paw inflammation. *Journal of Pharmacology and Toxicological Methods* 28(1): 15–19.

Gupta, U., Y. W. Jame, C. A. Campbell, A. J. Leyshon and W. Nicholaichuk. 1985. Boron toxicity and deficiency: A review. *Canadian Journal of Soil Science* 65(3): 381–409.

Gupta, P. K. and R. K. Varshney. 2000. The development and use of microsatellite markers for genetic analysis and plant breeding with emphasis on bread wheat. *Euphytica* 113: 163–85.

Gurung, S. and N. Skalko-Basnet. 2009. Wound healing properties of *Carica papaya* latex: *In vivo* evaluation in mice burn model. *Journal of Ethnopharmacology* 121(2): 338–41.

Hagel, J. M., E. C. Yeung and P. J. Facchini. 2008. Got milk? The secrate life of laticifers. *Trends in Plant Sciences* 13(12): 631–39.

Hama, E., S. Takumi, Y. Ogihara and K. Murai. 2004. Pistillody is caused by alterations to the class-B MADS-box gene expression pattern in alloplasmic wheats. *Planta* 218: 712–20.

Hamilton, R. A. 1954. A quantitative study of growth and fruiting in inbred and crossbred progenies from two Solo papaya strains. *Hawaii Agriculture Experimental Station Bulletin* 20: 16.

Hao, Ma., P. H. Moore, Z. Liu et al. 2004. High-density linkage mapping revealed suppression of recombination at the sex determination locus in papaya. *Genetics* 166: 419–36.

Hardisson, A., C. Rubio, A. Baez, M. M. Martin and R. Alvarez. 2001. Mineral composition of the papaya (*Carica papaya* variety Sunrise) from Tenerife Island. *European Food Research and Technology* 212: 175–81.

Harrington, J. F. 1972. Seed storage and longevity, *Seed Biology*, Vol. 3. New York: Academic Press, p. 145.

Hasan, M. F., T. M. M. Mahmud, J. Kadir, P. Ding and I. S. M. Zaidul. 2012. Sensitivity of *Colletotrichum gloeosporioides* to sodium bicarbonate on the development of anthracnose in papaya (*Carica papaya* L. cv. Frangi). *Australian Journal of Crop Science* 6(1): 17–22.

Hernandez, Y., M. G. Lobo and M. Gonzalez. 2006. Determination of vitamin c in tropical fruits: A comparative evaluation of methods. *Food Chemistry* 96: 654–64.

Hernandez, Y., M. G. Lobo and M. Gonzalez. 2009. Factors affecting sample extraction in the liquid chromatographic determination of organic acids in papaya and pineapple. *Food Chemistry* 114: 734–41.

Hine, R. B., O. V. Holtzmann and R. D. Raabe. 1965. Diseases of papaya (*Carica papaya* L.) in Hawaii. *Hawaii Agricultural Experiment Station Bulletin* 136: 1–25.

Hofmann, P. and A. M. Steiner. 1989. An updated list of recalcitrant seeds. *Landwirtschaftliche Forschung* 42: 310–23.

Hofmeyr, J. D. J. 1938. Genetical studies of *Carica papaya* L. *South African Department Agriculture Forest Science Bulletin* 187: 64.

Hofmeyr, J. D. J. 1939. Sex reversal in *Carica papaya* L. *South Africa Journal of Science* 26: 286–87.

Hofmeyr, J. D. J. 1942. Inheritance of dwarfness in *Carica papaya* L. *South African Department Agriculture Forest Science Bulletin* 45: 96–99.

Hofmeyr, J. D. J. 1945. Further studies of tetraploidy in *C. papaya* L. *South African Journal of Science* 41: 225–30.

Hofmeyr, J. D. J. 1953. Sex reversal as a means of solving breeding problems of *Carica papaya* L. *South African Journal of Science* 49: 228–32.

Hofmeyr, J. D. J. 1967. Some genetic and breeding aspects of *Carica papaya*. *Agronomy Tropical* 17: 345–51.

Horovitz, S. and H. Jimenez. 1967. Cruzamientosinterespecificos e intergenericos en Caricaceas y susimplicacionesfitotecnias. *Agronomy Tropical* 17: 353–59.

Huet, J., Y. Looze, K. Bartik, V. Raussens, R. Wintjens and P. Boussard. 2006. Structural characterization of the papaya cysteine proteinases at low pH. *Biochemical and Biophysical Research Communications* 341: 620–26.

Hunter, J. E. and R. K. Kunimoto. 1974. Dispersal of *Phytophthora palmivora* sporangia by wind-blown rain. *Phytopathology* 64: 202–06.

Husain, M. K., M. Anis and J. Shahzad. 2007. An *in vitro* propagation of Indian Kino (*Pterocarpus marsupium* Roxb.) using thidiazuron. *In vitro Cellular and Developmental Biology-Plant* 43: 59–64.

IBPGR. 1988. Descriptors for papaya. *International Board for Plant Genetic Resources*, Rome (Italy), 1–32.

Ighere Dickson, A., E. E. David, B. I. Temitope, A. A. Ahmed and M. Clement. 2012. Herbs used by the urhobo people in delta state Nigeria for the treatment of typhoid fever. *International Journal of Recent Scientific Research* 3(6): 478–81.

Imaga, N. A., Gbenle, G. O., Okochi, V. I. et al. 2010. Phytochemical and anti-oxidant constituents of *Carica papaya* and *Parquetina nigrescens* extracts. *Scientific Research Essays* 5(16): 2201–05.

Irtwange, S. V. 2006. Application of modified atmosphere packaging and related technology in postharvest handling of fresh fruits and vegetables. *Agriculture Engineering International* 4: 1–12.

Islam, R. and O. I. Joarder. 1996. Totipotency of *Carica papaya*. *Rice Biotechnology Quartarly* 26: 23.

Iyer, C. P. A. and R. M. Kurian. 2006. *High Density Planting in Tropical Fruits: Principles and Practice*. International Book Distributing Co, Delhi.

Iyer, C. P. A. and M. D. Subramanyam. 1981. Exploitation of heterosis in papaya *(Carica papaya* L.). *National Symposium on Tropical and Subtropical Fruit Crops*, Bangalore.

Iyer, C. P. A. and M. D. Subramanyam. 1984. Using of bridging species in interspecific hybridization in genus *Caricacae*. *Current Science* 53(24): 1300–01.

Iyer, C. P. A., M. D. Subramanyam and M. R. Dinesh. 1987. Interspecific hybridization in genus *Carica*. *IIHR Annual Reportt*, Bangalore, India.

Jaime, A., S. Teixeira da, R. Zinia et al. 2007. Papaya (*Carica papaya* L.) biology and biotechnology. *Tree, Forest Science and Biotechnology* 1(1): 47–73.

Jang, E. B. and D. M. Light. 1991. Behavioral responses of female oriental fruit flies to the odor of papayas at three ripeness stages in a laboratory flight tunnel (Diptera: Tephritidae). *Journal of Insect Behaviour* 4(6): 751–53.

Jayaraj, V., R. Suhanya, M. Vijayasarathy, M. Anandagopu and P. Anandagopu. 2009. Role of large hydrophobic residues in proteins. *Bioinformation* 3: 409–12.

Jesus de, O. N., J. P. X. de Freitas, J. L. L. Dantas and E. J. de Oliveira. 2012. Use of morpho-agronomic traits and DNA profiling for classification of genetic diversity in papaya. *Genetics and Molecular Resesarch* 11(4): 6646–62.

Jimenez, V. M., E. M. Newcomer and M. V. Gutierrez-Soto. 2014. Biology of the papaya plant. In *Genetics and Genomics of Papaya. Plant Genetics and Genomics: Crops and Models*, eds. R. Ming and P. H. Moore. New York: Springer Science + Business Media, 438.

Jobin-Decor, M. P., G. C. Graham, R. J. Hendry and R. A. Drew. 1997. RAPD and isozyme analysis of genetic relationships between *Carica papaya* and wild relatives. *Genetic Resources and Crop Evolution* 44(5): 471–77.

Jordan, M. and J. Velozo. 1996. Improvement of Somatic embryogenesis in highland- papaya cell suspensions. *Plant Cell, Tissue and Organ Culture* 44(3): 189–94.

Jordan, M. and J. Velozo. 1997. *In vitro* propagation of highland papayas (*C. pubescens* and *C. pentagona*). *Acta Horticulturae* 447: 103–06.

Juárez-Rojop, I. E., J. C. Diaz-Zagoya, J. L. Ble-Castillo et al. 2012. Hypoglycemic effect of *Carica papaya* leaves in streptozotocin-induced diabetic rats. *BMC Complementry and Alternative Medicine* 12: 236.

Kader, A. A. 2000. Postharvest technology research and information center. Department of Pomology, University of California, Davis, CA. Paw paw. Recommendations for Maintaining Postharvest Quality.

Kader, A. A. 2006. Papaya: Recommendations for maintaining postharvest quality. Postharvest Technology Research Information Center. Department of Plant Sciences, University of California.

Kader, A. A. and W. Płocharski. 1997. Fruit maturity, ripening, and quality relationships. *Acta Horticulturae* 485: 203–08.

Kader, A. A. and R. S. Rolle. 2004. The role of postharvest management in assuring the quality and safety of horticultural produce. *FAO Agricultural Support Systems Division* 152: 1010–365.

Kalleshwaraswamy, C. M. and N. K. K. Kumar. 2007. Transmission efficiency of papaya ringspot virus by three aphid species. *Phytopathology* 98: 541–46.

Kalleshwaraswamy, C. M., N. K. K. Kumar, A. Verghese, M. R. Dinesh, H. R. Ranganath and R. Venugopalan. 2007. Role of transient aphid vectors on the temporal spread of papaya ringspot virus in south India. *Acta Horticulturae* 740: 251–58.

Kamal Kumar, R., R. Amutha, S. Muthulaksmi, P. Mareeswari and W. B. Rani. 2007. Screening of dioecious papaya hybrids for papain yield and enzyme activity. *Research Journal of Agriculture Biological Sciences* 3: 447–49.

Kamphuis, I. G., K. H. Kalk, M. B. Swarte and J. Drenth. 1984. The structure of papain refined at 1.65- A resolution. *Journal of Molecular Biology* 179: 233–56.

Kays, J. S. 1997. *Postharvest Physiology of Perishable Plant Products*. Athens: University of Georgia.

Khade, S. W., B. F. Rodrigues and P. K. Sharma. 2010. Arbuscular mycorrhizal status and root phosphatase activities in vegetative *Carica papaya* L. varieties. *Acta Physiology Plant* 32(3): 565–74.

Kermanshai, R., B. E. McCarry, J. Rosenfeld, P. S. Summers, E. A. Weretilnyk and G. J. Sorger. 2001. Benzylisothiocyanate is the chief or sole anthelmintic in papaya seed extracts. *Phytochemistry* 57(3): 427–35.

Khan, S., A. P. Tyagi, A. Jokhan. 2002. Sex ratio in Hawaiian papaya (*Carica papaya* L.) variety Solo. *South Pacific Journal of Natural Science* 20: 22–24.

Khanna, N. and P. C. Panda. 2007. The effect of papain on tenderization and functional properties of spending hen meat cuts. *Indian Journal of Animal Research* 41: 55–58.

Khuspe, S. S. and S. D. Ugale. 1977. Floral biology of *Carica papaya* Linn. *Journal of Maharashtra Agricultural University* 2: 115–18.

Kim, J. K., M. S. Chu, S. J. Kim et al. 2010. Variation of glucosinolates in vegetable crops of *Brassica rapa L.* ssp. *Pekinensis*. *Food Chemistry* 119: 423–28.

Kim, M. S., P. H. Moore, F. Zee et al. 2002. Genetic and molecular characterization of *Carica papaya* L. *Genomes* 45: 503–12.

Kimmel, J. R. and E. L. Smith. 1954. Crystalline papain: I. Preparation, specificity and activation. *Journal of Biology and Chemistry* 207: 515–31.

Ko, W. H. 1982. Biological control of Phytophthora root rot of papaya. *Plant Disease* 66: 446–525.

Koffi, C. N. B., H. A. Diallo, J. Y. Kouadio, P. Kelly, A. G. Buddie and L. M. Tymo. 2010. Occurrence of *Pythium aphanidermatum* root and collar rot of papaya (*Carica papaya*) in Cote d'Ivoire. *Fruit, Vegetable and Cereal Science and Biotechnology* 4(1): 62–67.

Konno, K., C. Hirayama, M. Nakamura et al. 2004. Papain protects papaya trees from herbivorous insects: Role of cysteine proteases in latex. *Plant Journal* 37(3): 370–78.

Krishna, K. L., M. Paridhavi and J. A. Patel. 2008. Review on nutritional and pharmacological properties of papaya (*Carica papaya* L.). *Natural Product Radiance* 7: 364–73.

Kuang, Y. F. and Y. H. Chen. 2004. Induction of apoptosis in a non small cell human lung cancer cell line by isothiocyanates is associated with P53 and P21. *Food and Chemical Toxicology* 42: 1711–18.

Kumar, N. K. K., H. S. Singh and C. M. Kalleshwaraswamy. 2010. Aphid (Aphididae: Homoptera) vectors of *papaya ringspot virus* (PRSV), bionomics, transmission efficiency and factors contributing to epidemiology. *Acta Horticulturae* 851: 431–43.

Lai, C. C., S. D. Yeh and J. S. Yang. 2000. Enhancement of papaya axillary shoot proliferation *in vitro* by controlling the available ethylene. *Botanical Bulletin Academia Sinica* 41: 203–12.

Lai, C. W., Q. Yu, S. Hou et al. 2006. Analysis of papaya BAC end sequences reveals first insights into the organization of a fruit tree genome. *Molecular Genetics and Genomics* 276: 1–12.

Lange, A. H. 1961. Factors affecting sex changes in flowers of *Carica papaya* L. *Proceeding of American Society of Horticulture Sciences* 77: 252–64.

Lazan, H., Z. M. Ali and W. C. Sim. 1990. Retardation of ripening and development of water stress in papaya fruit seal-packaged with polyethylene film. *Acta Horticulturae* 269: 345–58.

Leal-Costa, M. V., M. Munhoz, P. E. Meissner Filho, F. Reinert and E. S. Tavares. 2010. Anatomia foliar de plantas transgênicas e nao transgenicas de *Carica papaya* L. (Caricaceae). *Acta Botanica Brasilica* 24: 595–97.

Leal, A. A., C. A. Mangolin, A. T. do Amaral Junior et al. 2010. Efficiency of RAPD versus SSR markers for determining genetic diversity among popcorn lines. *Genetics Molecular Research* 9: 9–18.

Lee, S. K. and A. A. Kader. 2000. Preharvest and postharvest factors influencing vitamin C content of horticultural crops. *Postharvest Biology and Technology* 20: 207–20.

Lee, W. T. 2003. Successive group-rearing of *Mallada basalis* with microcapsulated artificial diet and cost analysis. *Plant Protection Bulletin* 45: 45–52.

Lehane, R. 1996. Two-pronged assault on papaw virus. *Partners in Research for Development* 9: 32–38.

Li, F. and N. Xia. 2005. Population structure and genetic diversity of an endangered species, *Glyptostrobus pensilis* (Cupressaceae). *Botenical Bulletin Sinica* 46: 155–61.

Li, Z. Y., Y. Wang, W. T. Shen and P. Zhou. 2012. Content determination of benzyl glucosinolate and anti-cancer activity of its hydrolysis product in *Carica papaya* L. *Asian Pacific Journal of Tropical Medicines* 5(3): 231–33.

Lin, M., T. Kuo and C. Lin. 1998. Molecular cloning of cDNA encoding copper/zinc superoxide dismutase from papaya fruit and over expression in *Escherichia coli. Journal of Agriculture and Food Chemistry* 46(1): 344–48.

Litz, R. E. 1978. *In vitro* propagation of papaya. *Horticulture Science* 13: 241–42.

Litz, R. E. and R. A. Conover. 1978. Recent advances in papaya tissue culture. *Proceedings of The Florida State Horticultural Society* 91: 180–82.

Liu, C. Z., S. J. Murch, E. L. Demerdash and P. K. Saxena. 2003. Regeneration of the Egyptian medicinal plant *Artemisia judaica* L. *Plant Cell Reports* 21: 525–30.

Liu, Z., P. H. Moore, H. Ma et al. 2004. A primitive Y chromosome in papaya marks incipient sex chromosome evolution. *Nature* 427: 348–52.

Lo, K. C. 2002. Biological control of insect and mite pests on crops in Taiwan, review and prospection. *Formosan Entomology Special Publication* 3: 1–25.

Lohiya, N. K. and R. B. Goyal. 1992. Antifertility investigations on the crude chloroform extract of *Carica papaya* Linn. seeds in male albino rats. *Indian Journal of Experimental Biology* 30(11): 1051–55.

Lohiya, N. K., N. Pathak, P. K. Mistra, B. Maniovannan, S. S. Bhande, S. Panneerdoss and S. Sriram. 2005. Efficacy trial on the purified compounds of the seeds of *Carica papaya* for male contraception in albino rats. *Reproduction and Toxicology* 20(1): 135–48.

Lopes, M. C., R. C. Mascarini, B. M. de Silva, F. M. Florio and R. T. Basting. 2007. Effect of a papain-based gel for chemo-mechanical caries removal on dentin shearbond strength. *Journal of Dentistry for Children* 74: 93–97.

Louw, A. J. 1994. Papaya breeding programme. *Inlightings Bulletin Instittut Vir Tropiese en Subtropiese Gewasse* 266: 144.

Ma, H., P. H. Moore, Z. Liu et al. 2004. High-density linkage mapping revealed suppression of recombination at the sex determination locus in papaya. *Genetics* 166: 419–36.

Mabberley, D. J. 1998. *The Plant-Book: A Portable Dictionary of the Higher Plants*, 2nd edn., rev. printing. Cambridge, U.K: Cambridge Univ. Press.

Madej, T., K. J. Addess, J. H. Fong et al. 2012. MMDB: 3-D structures and macromolecular interactions. *Nucleic Acids Research* 40: D461–64.

Magdalita, P. M., S. W. Adkins, I. D. Godwin and R. A. Drew. 1996. An improved embryo-rescue protocol for a *Carica interspecific* hybrid. *Australian Journal of Botany* 44(3): 343–53.

Magdalita, P. M., R. A. Drew, I. D. Godwin and S. W. Adkins. 1998. An efficient interspecific hybridization protocol for *Carica papaya* L. x *Carica cauliflora* Jacq. *Australian Journal of Experimental Agriculture* 38(5): 523–30.

Magdalita, P. M., I. D. Godwin and R. A. Drew. 1997. Randomly amplified polymorphic DNA markers for a *Carica* interspecific hybrid. *Acta Horticulturae* 461: 133–40.
Magill, W., N. Deighton, H. W. Pritchard, E. E. Benson and B. A. Goodman. 1994. Physiological and biochemical studies of seed storage parameters in *Carica papaya*. *Proceeding of the Royal Society of Edinburgh* 102B: 439–42.
Mahmood, A. A., K. Sidik and I. Salmah. 2005. Wound healing activity of *Carica papaya* L. aqueous leaf extract in rats. *International Journal of Molecular Medicine and Advance Sciences* 1(4): 398–401.
Mahon, R. E., M. F. Bateson, D. A. Chamberlain, C. M. Higgins, R. A. Dres and J. L. Dale. 1996. Transformation of an Australian variety *Carica papaya* using microprojectile bombardment. *Australian Journal of Plant Physiology* 23(6): 679–85.
Mahouachi, J., A. Socorro and M. Talon. 2006. Responses of papaya seedlings (*Carica papaya* L.) to water stress and re-hydration: Growth, photosynthesis and mineral nutrient imbalance. *Plant Soil* 281(1): 137–46.
Maniyar, Y. and P. Bhixavatimath. 2012. Antihyperglycemic and hypolipidemic activities of aqueous extract of *Carica papaya* Linn. leaves in alloxan-induced diabetic rats. *Journal of Ayurveda and Integrative Medicines* 3(2): 70–74.
Mano, R., A. Ishida, Y. Ohya, H. Todoriki and S. Takishita. 2009. Dietary intervention with okinawan vegetables increased circulating endothelial progenitor cells in healthy young woman. *Atherosclerosis* 204: 544–48.
Manshardt, R. 2012. The Papaya in Hawaii presented at workshop History of Hawaiian Pomology held during September 25, 2011 at the ASHS Conference, Waikoloa, HI. *Hortscience* 47(10): 1399–404.
Manshardt, R. M. 1992. Papaya. In *Biotechnology of Perennial Fruit Crops*, eds. F. A. Hammerschlag and R. E. Litz. Wallingford, U.K: CAB International, 489–511.
Manshardt, R. M. and R. A. Drew. 1998. Biotechnology of papaya. *Acta Horticulturae* 461: 65–73.
Mantok, C. 2005. Multiple usage of green papaya in healing a Tao garden. Tao garden health spa and resort, Thailand. Retrieved from: www.tao.garden.com.
Marelli de Souza, L., K. S. Ferreira, J. B. P. Chaves and D. S. Teixeira. 2008. L-ascorbic acid, β-carotene and lycopene content in papaya fruits (*Carica papaya*) with or without physiological skin freckles. *Scientia Agricola (Piracicaba, Brazil)* 65: 246–50.
Marler, T. E. 2011. Growth responses to wind differ among papaya roots, leaves and stems. *Horticulture Science* 46(8): 1105–09.
Marler, T. E. and H. M. Discekici. 1997. Root development of Red Lady papaya plants grown on a hillside. *Plant Soil* 195(1): 37–42.
Marler, T. E., A. P. George, R. J. Nissen and P. C. Andersen. 1994. Miscellaneous tropical fruits. In *Handbook of Environmental Physiology of Fruit Crops*, eds. B. Scheffer and P. C. Andersen. Vol II, Sub-tropical and tropical Crops. CRC Press, Boca Raton, 199–224.
Marler, T. E. and M. V. Mickelbart. 1998. Drought, leaf gas exchange, and chlorophyll fluorescence of field-grown papaya. *Journal of American Society for Horticulture Sciences* 123(4): 714–18.
Martins, D. D. S., M. P. Culik and V. D. S. Wolff. 2004. New record of scale insects (Hemiptera: Coccoidea) as pests of papaya in Brazil. *Neotropical Entomology* 33(5): 655–57.
Martins, D. J. and S. D. Johnson. 2009. Distance and quality of natural habitat influence hawkmoth pollination of cultivated papaya. *International Journal of Tropical Insect Sciences* 29(3): 114–23.
Matsuura, F. C. A. U., M. I. D. S. Folegatti, R. L. Cardoso and D. C. Ferreira. 2004. Sensory acceptance of mixed nectar of papaya, passion fruit and acerola. *Scientia Agricola (Piracicaba, Brazil)* 61: 604–08.
Mc Candless, L. 1997. Genetic engineering performs miracles with plants. *Cornell Focus* 6(1): 20–24.

Medina, J. D. L. C., G. V. Gutierrez and H. S. Garcia. 2003. Pawpaw: Post-harvest Operation. Edited by AGSI/FAO: Danilo Mejía, PhD, AGST, FAO (Technical), p. 70.

Mehdipour, S., N. Yasa, G. Dehghan et al. 2006. Antioxidant potentials of Iranian *Carica papaya* juice *in vitro* and *in vivo* are comparable to alpha-tocopherol. *Phototherapy Research* 20(7): 591–94.

Mekako, H. V. and H. Y. Nakasone. 1975. Interspecific hybridization in papaya. *Journal of the American Society for Horticultural Science* 100: 237–42.

Menard, R., H. E. Khouri, C. Plouffe et al. 1990. A protein engineering study of the role of aspartate 158 in the catalytic mechanism of papain. *Biochemistry* 29: 6706–13.

Mendoza, E. M. T. 2007. Development of functional foods in the Philippines. *Food Science and Technology Research* 13: 179–86.

Miller, D. R. and G. L. Miller. 2002. Redescription of *Paracoccus marginatus* Williams and Granara de Willink (Hemiptera: Coccoidea: Pseudococcidae), including descriptions of the immature stages and adult male. *Proceedings of the Entomological Society of Washington*, 104: 1–23.

Ming, R., A. Bendahmane and S. S. Renner. 2011. Sex chromosomes in land plants. *Annual Review of Plant Biology* 62: 485–514.

Ming, R., S. Hou, Y. Feng et al. 2010. The draft genome of the transgenic tropical fruit tree papaya (*Carica papaya* Linnaeus). *Nature* 452: 991–96.

Ming, R., Q. Yu, A. Blas, C. Chen, N. Jong-Kuk and P. H. Moore. 2008. Genomics of papaya, a common source of vitamins in the tropics. *Genomics of Tropical Crop Plants, Plant Genetics and Genomics: Crops and Models* 1: 405–20.

Ming, R, Q. Yu and P. H. Moore. 2007. Sex determination in papaya. *Seminar on Cellular and Developmental Biology* 18: 401–08.

Ming, R. and P. H. Moore. 2014. Genetics and genomics of papaya. *Plant Genetics and Genomics: Crops and Models,* 10, Springer Science + Business Media, Xiii, 438p, New York.

Ming, R., P. H. Moore, F. Zee, C. A. Abbey, H. Ma and A. H. Paterson. 2001. Construction and characterization of a papaya BAC library as a foundation for molecular dissection of a tree-fruit genome. *Theoretical Applied Genetics* 102: 892–99.

Mitchel, R. E., M. I. Claiken and E. L. J. Smith. 1970. The complete amino acid sequence of papain. *Journal of Biology and Chemistry* 245: 3485–92.

Mitchell, C. H. and P. K. Diggle. 2005. The evolution of unisexual flowers: Morphological and functional convergence results from diverse developmental transitions. *American Journal of Botany* 92: 1068–76.

Miyoshi, N., K. Uchida, T. Osawa and Y. Nakamura. 2007. Selective cytotoxicity of benzyl isothiocyanate in the proliferating fibroblastoid cells. *International Journal of Cancer* 120(3): 484–92.

Moore, P. H. and R. Ming. 2008. Papaya genome: A model for tropical fruit trees and beyond. *Tropical Plant Biology* 1: 179–80.

Momenzadeh, L., A. Zomorodian, D. Mowla. 2010. Experimental and theoretical investigation of shelled corn drying in a microwave assisted fluidized bed dryer using Artificial Neural Network. *Food and Bioproducts Processing* 89(1): 15–21.

Monti, R., C. A. Basilio, H. C. Trevisan and J. Contiero. 2000. Purification of papain from fresh latex of *Carica papaya*. *Brazilian Archives Biology Technology* 43: 501–07.

Mora-Aguilera, G., D. Teliz, C. L. Campbell and C. Avilla. 1992. Temporal and spatial development of papaya ringspot in Veracruz, Mexico. *Journal of Phytopathology* 136: 27–36.

Morales-Payan, J. P. and W. M. Stall. 2005. Papaya transplant growth as affected by 5-aminolevulinic acid and nitrogen fertilization. *Proceeding on Florida State Horticulture Society* 118: 263–65.

Morgante, M. and A. M. Olivieri. 1993. PCR amplified microsatellites in plant genetics. *Plant Journal* 3: 175–82.

Morimoto, C., N. H. Dang and N. Y. S. Dang. 2008. Cancer prevention and treating composition for preventing, ameliorating, or treating solid cancers, e.g. lung, or blood cancers, e.g. lymphoma, comprises components extracted from brewing papaya. *Patent number-WO2006004226-A1; EP1778262- A1;JP2008505887-W; US2008069907-A1.*

Morshidi, M. 1998. Genetic control of isozymes in *Carica papaya* L. *Euphytica* 103(1): 89–94.

Morton, J. 1987. Papaya. In *Fruits of Warm Climates*, eds. J. F. Morton and F. L. Miami, 336–346. The New Crop Resource Online Programme, Purdue University. Available at http://www.hort.purdue.edu/newcrop/morton/papaya_ars.html

Mossler, M. A. and O. N. Nesheim. 2002. Florida crop pest management profile: Papaya. *FL* 352: 392–472.

Muller, R. and P. S. Gooch. 1982. Organic amendments in nematode control. An examination of literature. *Nematropica* 12: 319–26.

Muniappan, R., D. E. M. Eyerdirk, F. M. S. Engebau, D. D. B. Erringer and G. V. P. Reddy. 2006. Classical biological control of the papaya mealybug, *Paracoccus marginatus* (hemiptera: pseudococcidae) in the Republic of Palau. *Florida Entomologist* 89(2): 212–17.

Murashige, T. and F. Skoog. 1962. A revised medium for rapid growth and bioassays with tobacco tissue cultures. *Physiology Plant* 15: 473–97.

Nagappan, G. and N. Surugau. 2011. Benzyl glucosinolate hydrolysis products in papaya (*Carica papaya*). In Conference held at Malaysia University Institutional Repository, Malaysia. Available at http://eprints.ums.edu.my/8615/.

Nakamura, Y., M. Yoshimoto, Y. Murata et al. 2007. Papaya seed represents a rich source of biologically active isothiocyanate. *Journal of Agricultural and Food Chemistry* 55: 4407–13.

Nakasone, H. Y. 1967. Papaya breeding in Hawaii. *Agronomy Tropical* 17: 391–99.

Nakasone, H. Y. 1986. Papaya. In *CRC Handbook of Fruit Set and Development*, ed. S. P. Monselise, 277–301. Boca Raton , Florida: CRC Press.

Nakasone, H. Y. and R. E. Paull. 1998. *Tropical Fruits.* Wallingford, UK: CAB International.

Nakasone, H. Y. and R. E. Paull. 1999. *Tropical Fruits.* Wallingford, UK: CABI Publishing.

Nakasone, H. Y. and W. B. Storey. 1955. Studies on the inheritance of fruiting height of *Carica papaya. Proceeding for American Society of Horticulture Science* 66: 168–82.

Nandi, A. K. and B. C. Mazumdar. 1990. Biochemical differences between male and female papaya *(Carica papaya* L.) trees in respect to total RNA and the histone protein level. *Indian Biologist* 22(1): 47–50.

Naturlande. 2000. Organic farming in the tropics and subtropics (exemplary description of 20 crops). Naturlande. V– 1st edition. Available at http://www.naturland.de/fileadmin/MDB/documents/Publication/English/papaya.pdf.

Nayak, S. B., P. L. Pereira and D. Maharaj. 2007. Wound healing activity of *Carica papaya* L. in experimentally induced diabetic rats. *Indian Journal of Experimental Biology* 45(8): 739–43

Nayak, B. S., R. Ramdeen, A. Adogwa, A. Ramsubhag and J. R. Marshall. 2012. Wound-healing potential of an ethanol extract of *Carica papaya* (Caricaceae) seeds. *International Wound Journal* 9(6): 650–55.

Nelson, S. 2008. Phytophthora blight of papaya. *College of Tropical Agriculture and Human Resources* 53: 1–7.

Nelson, S. 2012a. Boron deficiency of papaya. Honolulu (HI): University of Hawaii. Plant Disease; PD-91:4.

Nelson, S. 2012b. Bacterial wilt of papaya: Wilting, chlorosis and necrosis. Available at https://www.flickr.com/photos/scotnelson/sets/72157634416071328/.

Neupane, K. R., U. T. Mukatira, C. Kato and J. I. Stiles. 1998. Cloning and characterization of fruit-expressed ACC synthase and ACC oxidase from papaya (*Carica papaya* L.). *Acta Horticulturae* 461: 329–37.

Nguyen, T. T., P. N. Shaw, M. O. Parat and A. K. Hewavitharana. 2013. Anticancer activity of *Carica papaya*: A review. *Molecular, Nutritional and Food Research* 57(1): 153–64.

NHB (National Horticulture Board). 2002. Papaya diseases. Available at http://nhb.gov.in/fruits/papaya/pap002.pdf.

Nishijima, W. T. 1999. Diseases of papaya. *American Phytopathological Society*. Available at www.apsnet.org/online/common/names/papaya.asp.

Nishina, M., F. Zee, R. Ebesu et al. 2000. Papaya production in Hawaii. Fruits and nuts. *College of Tropical Agriculture and Human Resources* 3: 1–8.

Nishina, M. S. 1991. Bumpy fruit of papaya as related to boron deficiency. County Extension Agent, Hawaii Cooperative Extension Service Hawaii Institute of Tropical Agriculture and Human Resources, University of Hawaii at Manoa Commodity Fact Sheet, Fruit Pa-4(B).

Noa-Carrazana, J. C., D. Gonzalez-de-Leon, B. S. Ruiz-Castro, D. Pinero and L. Silva-Rosales. 2006. Distribution of papaya ringspot virus and papaya mosaic virus in papaya plants (*Carica papaya*) in Mexico. *Plant Diseases* 90: 1004–11.

Nwofia, G. E., P. Ogimelukwe and C. Eji. 2012. Chemical composition of leaves, fruit pulp and seed in some morphotypes of *C. papaya* L. morphotypes. *International Journal of Medicinal and Aromatic Plants* 2: 200–06.

Oboh, G., A. A. Olabiyi, A. J. Akinyemi and A. O. Ademiluyi. 2013. Inhibition of key enzymes linked to type 2 diabetes and sodium nitroprusside-induced lipid peroxidation in rat pancreas by water-extractable phytochemicals from unripe pawpaw fruit (*Carica papaya*). *Journal of Basic Clinical Physiology and Pharmacology* 30: 1–14.

Ocampo, J., C. G. Eeckenbrugge, S. Bruyere, L. D. Bellaire and P. Ollitrault. 2006. Organization of morphological and genetic diversity of Caribbean and Venezuelan papaya germplasm. *Fruits* 61: 25–37.

Ocampo Perez, J., G. C. d'Eeckenbrugge, A. M. Risterucci, D. D. Bier and P. Ollitrault. 2007. Papaya genetic diversity assessed with microsatellite markers in germplasm from the Caribbean Region. *Acta Horticulturae* 740: 93–101.

Odani, S., Y. Yokokawa, H. Takeda, S. Abe and S. Odani. 1996. The primary structure and characterization of carbohydrate chains of the extracellular glycoprotein proteinase inhibitor from latex of *Carica papaya*. *European Journal of Biochemistry* 241(1): 77–82.

Odu, E. A., O. Adedeji and A. Adebowale. 2006. Occurrence of hermaphroditic plants of *Carica papaya* L. (Caricaceae) in Southwestern Nigeria. *Journal of Plant Sciences* 1: 254–63.

Oduola, T., I. Bello, T. Idowu, G. Avwioro, G. Adeosun and L. H. Olatubosun. 2010. Histopathological changes in Wistar albino rats exposed to aqueous extract of unripe *Carica papaya*. *North American Journal of Medical Sciences* 2: 234–37.

OECD (Organisation for the Economic Cooperation and Development). 2005. Consensus document on the biology of papaya (*Carica papaya*). *Series on Harmonisation of Regulatory Oversight in Biotechnology*, OECD, Environment, Health and Safety publications, Directorate, Paris, 33. Available at http.//appli1.oecd.org/olis/2005doc.nsf/43bb6130e5e86e5fc12569fa005d004c/lelale18daeflcb9c12570a1004f93bd/$FILE/JT00192446.pdf.

OECD (Organisation for the Economic Cooperation and Development). 2008. Consensus document on compositional considerations for new varieties of tomato: Key food and feed nutrients, toxicants and allergens. In *Series on the Safety of Novel Foods and Feeds*, OECD Environment Directorate, Paris, 17.

OGTR (Office of the Gene Technology Regulator, Australia). 2003. The Biology and ecology of papaya (paw paw), *Carica papaya* L., in Australia. Office of the Gene Technology Regulator, Government of Australia.

OGTR (Office of the Gene Technology Regulator, Australia). 2008. The biology of *Carica papaya* L. (papaya, papaw, paw paw), Version 2: February 2008, Australian Government, Dpt. of Health and Ageing, OGTR. Available at http://www.ogtr.gov.au/internet/ogtr/publishing.nsf/Content/papaya/$FILE/biologypapaya08.pdf.

O'Hare, P. 1993. *Growing Papaya in South Queensland.* Brisbane, Queensland: Queensland Government Department of Primary Industries.

Okiniyi, J. A. O., T. A. Ogunlesi, O. A. Oyelami and L. A. Adeyemi. 2007. Effectiveness of dried *Carica papaya* against human intestinal parasitosis: A pilot study. *Journal of Medicinal Food* 10: 194–96.

Okoko, T. and D. Ere. 2012. Reduction of hydrogen peroxide-induced erythrocyte damage by *Carica papaya* leaf extract. *Asian Pacific Journal of Tropical Biomedicine* 2(6): 449–53.

Oliveira, E. J., J. L. L. Dantas, M. S. Castellen and M. D. Machado. 2008. Identificacao de microssatelites para o mamoeiro por meio da exploracao do banco de dados de DNA. *The Revista Brasileira de Fruticultura* 30: 841–45.

Oliveira, E. J., J. G. Padua, M. I. Zucchi, R. Vencovsky and M. L. C. Vieira. 2006. Origin, evolution and genome distribution of microsatellites. *Genetics and Molecular Biology* 29: 294–307.

Oloyede, O. I. 2005. Chemical profile of unripe pulp of *Carica papaya. Pakistan Journal of Nutrition* 4(6): 379–81.

Onibon, V. O., F. O. Abulude and L. O. Lawal. 2007. Nutritional and antinutritional composition of some Nigerian fruits. *Journal of Food Technology* 5: 120–22.

Otsuki, N., N. H. Dang, E. Kumagai, A. Kondo, S. Iwata and C. Morimoto. 2010. Aqueous extract of *Carica papaya* leaves exhibits anti-tumor activity and immunomodulatory effects. *Journal of Ethnopharmacology* 127(3): 760–67.

Owoyele, B. V., O. M. Adebukola, A. A. Funmilayo and A. O. Soladoye. 2008. Anti-inflammatory activities of ethanolic extract of *Carica papaya* leaves. *Inflammopharmacology* 16(4): 168–73.

Pandey, R. M. and S. P. Singh. 1988. Field performance of *in vitro* raised papaya plants. *Indian Journal of Horticulture* 45: 1–7.

Pandey, V. P., S. Singh, R. Singh and U. N. Dwivedi. 2012. Purification and characterization of peroxidase from papaya (*Carica papaya*) fruit. *Application of Biochemistry and Biotechnology* 167(2): 367–76.

Panjaitan, S. B., M. A. Aziz, A. A. Rashid and N. M. Saleh. 2007. *In vitro* plantlet regeneration from shoot tip of field-grown hermaphrodite papaya (*Carica papaya* L. cv. Eksotika). *International Journal of Agriculture and Biology* 6: 827–32.

Parasnis, A. S., V. S. Gupta, S. A. Tamhankar and P. K. Ranjekar. 2000. A highly reliable sex diagnostic PCR assay for mass screening of papaya seedlings. *Molecular Breeding* 6: 337–44.

Pares, J., C. Basso and D. Jauregui. 2002. Momento de antesis, dehiscencia de anteras y receptividad estigmática en flores de lechosa (*Carica papaya* L.) cv. Cartagena Amarilla. *Bioagrology* 14: 17–24.

Pares-Martinez, J., R. Linarez, M. Arizaleta and L. Melendez. 2004. Aspectos de la biologia floral en lechosa (*Carica papaya* L.) cv. Cartagena Roja, en el estado Lara, Venezuela. *Revista de la Facultad de Agronomia (LUZ)* 21: 116–25.

Paterson, A., P. Felker, S. Hubbell and R. Ming. 2008. The fruits of tropical plant genomics. *Tropical Plant Biology* 1(1): 3–19.

Paull, R. E. 1990. Postharvest heat treatments and fruit ripening. *Post-Harvest News Information* 1: 355–63.

Paull, R. E. 1993. Pineaple and Papaya. In *Biochemistry of Fruit Ripening*, eds. G. B. Symour, J. E. Taylor and G. A. Tucker, New York: Chapman and Hall, 291–311.

Paull, R. E., K. Cross and Y. Qiu. 1999. Changes in papaya cell walls during fruit ripening. *Postharvest Biology and Technology* 16: 79–89.

Paull, R. E., W. Nishijima, R. Marcelino and C. Cavaletto. 1997. Postharvest handling and losses during marketing of papaya (*Carica papaya* L.). *Postharvest Biology and Technology* 11: 165–79.

Paz, L. and C. Vazquez-Yanes. 1998. Comparative seed ecophysiology of wild and cultivated *Carica papaya* trees from a tropical rain forest region in Mexico. *Tree Physiology* 18(4): 277–80.

Peleg, M. and L. Gomez-Brito. 1975. The red component of the external color as a maturity index of papaya fruits. *Journal of Food Sciences* 40: 1105–06.

Pena, J. E. and F. A. Johnson. 2006. Insect management in papaya. *The Institute of Food and Agricultural Sciences (IFAS)*, University of Florida ENY-414:1–4.

Perera, M. R., R. D. F. Vargas and M. G. K. Jones. 2008. First record of infection of papaya trees with root-knot nematode (*Meloidogyne javanica*) in Australia. *Australian Plant Disease Notes* 3: 87–88.

Pesante, A. 2003. Market outlook report: Fresh papaya, Honolulu HI. Available at http://hawaii.gov/hdoa/add/researchandoutlookreports/papaya%20outlook%20report.pdf.

Phuangrat, B., N. Phironrit, A. Son-ong et al. 2013. Histological and morphological studies of pollen grains from elongata, reduced elongata and staminate flowers in *Carica papaya* L. *Tropical Plant Biology* 6(4): 210–16.

Picha, D. 2006. Horticultural crop quality characteristics important in international trade. *Acta Horticulturae* 712: 423–26.

Piva, E., F. A. Ogliari, R. R. D. Moraes, F. Cora and L. C. H. S. Sobrinho. 2008. Papain-based gel for biochemical caries removal: Influence on microtensile bond strength to dentin. *Brazilian Oral Research* 22: 364–70.

Porter, B. W., Y. J. Zhu, D. T. Webb and D. A. Christopher. 2009. Novel thigmomorphogenetic responses in *Carica papaya*: Touch decreases anthocyanin levels and stimulates petiole cork outgrowths. *Annals of Botany* 103(6): 847–58.

Prakash, A. and A. P. Dikshit. 1963. Studies on storage of papaya pollen. *Journal of Scientific Research BHU* 14(1): 52–55.

Prakash, J., S. P. Das, T. Bhattacharjee and N. P. Singh. 2014. Studies on effect of pollarding in papaya. *Indian Journal of Horticulture* 71(3): 419–20.

Prakash, J. and A. K. Singh. 2006. Potential and prospects of protected cultivation of fruit crops in India. *National Seminar on Protected Cultivation of Horticultural Crops and Value Addition*, Allahbad (UP), India, 145–48.

Punja, Z. K., S. Rose and R. Yip. 2002. Biocontrol agents and composts suppress Fusarium and Pythium root rots on greenhouse cucumbers. *International Organisation for Biological and Integrated Control of Noxious Animals and Plants* (OIBC/OILB), Vol. 25, pp. 93–96. West Palaearctic Regional Section (WPRS/SROP), Dijon, France.

Purohit, A. G. 1980. Effect of supplementary pollination on seed production in gynodioecious papaya var. Coorg honey dew. *Progressive Horticulture* 11: 63–66.

Puwastien, P., B. Burlingame, M. Raroengwichit and P. Sungpuag. 2000. *ASEAN Food Composition Tables of Nutrition*. Thailand: Mahidol University.

Quinta, M. E. G. and R. E. Paull. 1993. Mechanical injury during postharvest handling of 'Solo' papaya fruit. *Journal of American Society for Horticulture Sciences* 118: 618–22.

Rajbhar, P. Y., G. Singh and M. Lal. 2010. Effect of N, P, K and spacing on growth and yield of papaya (*Carica papaya* L.) cv. Pant papaya. *Acta Horticullturae* 851: 425–28.

Rajeevan, M. S. and R. M. Pandey. 1983. Propagation of papaya through tissue culture. *Acta Horticulturae* 137: 131–39.

Rallo, P., G. Dorado and A. Martin. 2000. Development of simple sequence repeats (SSRs) in olive tree (*Olea europaea* L.). *Theoretical Applied Genetics* 101: 984–89.

Ram, M. 1981. Pusa 1–15, An outstanding papaya. *Indian Horticulture* 26: 21–22.

Ram, M. 1982. Papaya improvement through selection and breeding technique. *Punjab Horticulture Journal* 22: 8–14.

Ram, M. 1983. Some aspects of genetics, cytogenetics and breeding of papaya. *South Indian Horticulture* 31: 34–43 (30th Year Commemorative Issue).

Ram, M. 1984a. Promising varieties of papaya. *Proceeding for Papaya and Papain Production* Seminar. Coimbatore, India.

Ram, M. 1984b. Beware of bud and fruit stalk rot of papaya. *Indian Horticulture* 29: 19–20.

Ram, M. 1986a. For pure papaya seed. *Intensive Agriculture* 23: 5–8.

Ram, M. 1986b. Seed production in papaya. *Punjab Horticulture Journal* 25: 95–101.

Ram, M. 1992. Utilization of genetic diversity of Indian papaya. *Proceeding for Production and Utilization of Papaya Seminar.* Coimbatore, India.

Ram, M. 1993. Improvement of papaya. In *Advances in Horticulture*, Vol. 1, eds. K. L. Chadha and O. P. Pareek. New Delhi: Malhotra Publishing House, pp. 383–97.

Ram, M. 1995. Papaya seed production under controlled pollination and isolation. *Seed Research* 23(2): 98–101.

Ram, M. 1996. Papaya. In *A Textbook on Pomology: Tropical Fruits,* ed. T. K. Chattopadhyay, India, Ludhiana: Kalyani Publishers, 113–40.

Ram, M. 2005. *Papaya*. Directorate of information and publication of agriculture, ICAR, KAB-I, Pusa New Delhi, India.

Ram, M. and P. K. Majumder. 1981. Dwarf mutant of papaya (*Carica papaya* L.) induced by gamma rays. *Journal of Nuclear Agriculture Biology* 10: 72–74.

Ram, M. and P. K. Majumder. 1984. Papaya. *Directory of Germplasm Collection, Tropical Fruits*. Netherland: IBPGR.

Ram, M. and P. K. Majumder. 1988. High density orcharding in papaya. *Journal of Institute for Agriculture and Animal Sciences* 9: 115–17.

Ram, M. and P. K. Majumder. 1992. Genetic divergence in papaya. *Proceedings for Production and Utilization of Papaya Seminar.* Coimbatore, India.

Ram, M., P. K. Majumder and B. N. Singh. 1985a. Genetics of sex reversing male in papaya. *Indian Journal of Horticulture* 42: 63–65.

Ram, M., P. K. Majumder and R. N. Singh. 1985b. Papaya germplasm collection in India. *IBPGR Newsletters* 9: 6–7.

Ram, M., P. K. Majumder, B. N. Singh and A. Akhtar. 1999. Study on hybrid vigour in papaya (*carica papaya* L.) varieties. *Indian Journal of Horticulture* 56(4): 295–98.

Ram, M. and A. K Pandey. 1990. Reducing cost of papaya cultivation through intercropping with tobacco. *New Frontiers in Horticulture Seminar* Banglore, India.

Ram, M. and P. K. Ray. 1992. Influence of fruiting season on seed production of papaya in north Bihar. *All India Seed Seminar on Recent Advances and Future Strategies in Seed Science and Technology*, Nauni, Solan, India.

Ram, M. and S. Srivastava. 1984. Mutagenesis in papaya. *National seminar on papaya and papain production*, TNAU Coimbatore, India.

Ramos, H. C. C., M. G. Pereira, F. F. Silva et al. 2011. Genetic characterization of papaya plants (*Carica papaya* L.) derived from the first backcross generation. *Genetics and Molecular Research* 10(1): 393–403.

Rana, S. and R. Sah. 2010. *Study of the effect of protected environment on pan evaporation and water requirement of different vegetables.* Dissertation, GBPUA&T., Pantnagar, India.

Rawal, R. D. 2010. Fungal diseases of papaya and their management. *Acta Horticulturae* 851: 443–46.

Ray, P. K. 2002. *Breeding Tropical and Subtropical Fruits.* Narosa Publishing House, Daryaganj, New Delhi, India.

Razafindraibe, M., A. R. Kuhlman, H. Rabarison et al. 2013. Medicinal plants used by women from Agnalazaha littoral forest (South Eastern Madagascar). *Journal of Ethnobiology and Ethnomedicine* 9: 73.

Reddy, S. R., R. B. Krishna and K. J. Reddy. 2012. Sex determination of papaya (*Carica papaya)* at seedling stage through RAPD Markers. *Research in Biotechnology* 3(1): 21–28.

Reiger, M. 2006. Papaya (*Carica papaya*). In *Introduction to Fruit Crops,* ed. M. Rieger, New York: Haworth Food and Agricultural Products Press, 301–310.

Reuveni, O., D. R. Shlesuiger and U. Lavi. 1990. *In vitro* clonal propagation of dioecious *Carica papaya* L. *Plant Cell, Tissue and Organ Culture* 20(1): 41–46.

Rex, A. and A. Rivera. 2005. Guide to papaya growing and marketing (online). *Agronomist, Agricultural Consultant*, 30 Lapu Street, General Santos City.

Rayes, M. N., A. Perez and J. Cueveas. 1980. Detecting endogenous growth regulators of the sarcotesta, sclerotesta, endosperm and embryo by paper chromatography in fresh and aged seed of two varieties of papaya. *Journal of the Agricultural University of Puerto Rico* 15: 164–72.

Reyes, O. S. and A. C. Fermin. 2003. Terrestrial leaf meals or freshwater aquatic fern as potential feed ingredients for farmed abalone *Haliotis asinina* (Linnaeus 1758). *Aquaculture Research* 34: 593–99.

Rimberia, F. K., S. Adaniya, Y. Ishimine and T. Etoh. 2009. Variation in ploidy and morphology among anther-derived papaya plants. *Acta Horticultrae* 829: 375–81.

Robert, A. M., L. Dann and G. Lown. 1974. The specificity of the S'1 sub-site of papain. *Journal of Biochemistry* 141: 495–501.

Rodrigues, S., M. Da Cunha, J. A. Ventura and P. Fernandes. 2009. Effects of the papaya meleira virus on papaya latex structure and composition. *Plant Cell Reports* 28(5): 861–71.

Rodriguez, M. C. and V. Galan. 1995. Preliminary study of paclobutrazol (PP333) effects on greenhouse papaya (*Carica papaya* l.) in the Canary Islands. *Acta Horticulturae* 370: 167–72.

Rodriguez, J., P. Rodriguez, M. E. Gonzalez and P. Martinez-Gomez. 2010. Molecular characterization of Cuban endemism *Carica cubensis* Solms using random amplified polymorphic DNA (RAPD) markers. *Agricultural Sciences* 1(2): 95–101.

Rodriquez-Pastor, M. C., G. V. Sauco and H. M. Romero. 1990. Evaluation of papaya antogeny. *Fruits (Peris)* 45(4): 387–91

Sagar, S. B., H. C. Parmar and V. B. Darji. 2012. Economics of production of papaya in middle Gujarat region of Gujarat, India. *Globle Journal of Biology, Agriculture and Health Sciences* 1(2): 10–17.

Saha, M. 2007. *In-vitro* propagation of precociously germinated seedlings of *Carica papaya* L. variety Madhubindu. *Bionano Frontier* 1(1): 55–59.

Sahu, P. R., R. Zhang, S. Batra, Y. Shi and S. K. Srivastava. 2009. Benzyl isothiocyanate-mediated generation of reactive oxygen species causes cell cycle arrest and induces apoptosis via activation of MAPK in human pancreatic cancer cells. *Carcinogenesis* 30: 1744–53.

Salomao, A. N. and R. C. Mundim. 2000. Germination of papaya seed in response to desiccation, exposure to subzero temperatures, and gibberellic acid. *Horticulture Science* 35(5): 904–06.

Samson, J. A. 1986. *Tropical Fruits*, 2nd edn., England: Longman Scientific and Technical.

Sankat, C. K. and R. Maharaj. 2001. Papaya. In *Postharvest Physiology and Storage of Tropical and Subtropical Fruits*, ed. S. Mitra, 167–185. India, West Bengal: CAB International.

Santos, S. C., C. Ruggiero, C. L. S. P. Silva and E. G. M. Lemos. 2003. A microsatellite library for *Carica papaya* L. cv. Sunrise Solo. *The Revista Brasileira de Fruticultura* 25: 263–67.

Saran, P. L. 2010. Screening of papaya cultivars under Doon Valley conditions. *Pantnagar Journal of Research* 8(2): 246–47.

Saran, P. L. and R. Choudhary. 2013. Drug bioavailability and traditional medicaments of commercially available papaya—a review. *African Journal of Agriculture Research* 8(25): 3216–23.

Saran, P. L., R. Choudhary and I. S. Solanki. 2013a. Micro-irrigation for sustainable papaya cultivation. Smarika, Rajya Stariya Sangoshti evm Perdarshni. *Uttarakhand me Suksham Sichai Padati- Chunotia evm Sambhavnaye* 33–34.

Saran, P. L., R. Choudhary and I. S. Solanki. 2014b. New papaya selections and relationship studies using morphological and molecular markers. In *An International Event on Horticulture for Inclusive Growth,* eds. K. L. Chadha, S. K. Singh, M. Srivastav and T. K. Behera. Tamil Nadu, India: Tamil Nadu Agricultural University Coimbatore, p. 154.

Saran, P. L., R. Choudhary and I. S. Solanki. 2014c. Papaya: Agri-hort system. *Agriculture Today* 17(2): 40–41.

Saran, P. L., R. Choudhary, I. S. Solanki and G. Devi. 2014d. Traditional medicaments through Papaya in North Eastern Plains Zone of India. In: *An International Event on Horticulture for Inclusive Growth,* eds. K. L. Chadha, S. K. Singh, M. Srivastav and T. K. Behera. Tamil Nadu, India: Tamil Nadu Agricultural University Coimbatore, p. 28.

Saran, P. L., R. Choudhary, I. S. Solanki and G. Devi. 2015. Traditional medicaments by papaya among the tribal communities in North Eastern plains zone of India. *Indian Journal of Traditional Knowledge* 14(1): (In Press).

Saran, P. L., R. Choudhary, I. S. Solanki and P. R. Kumar. 2014a. New physiological disorders in papaya (*Carica papaya* L.). *African Journal of Biotechnology* 13(4): 574–80.

Saran P. L., R. Choudhary, I. S. Solanki, P. Patil and S. Kumar. 2015. Genetic variability and relationship studies of new Indian papaya (*Carica papaya* L.) germplasm using morphological and molecular markers. *Turkish Journal of Agriculture and Forestry* 39(3): (In Press).

Saran, P. L., R. Choudhary, I. S. Solanki and K. Singh. 2013b. New bottlenecks in seed production of papaya under North Eastern Plains Zone. New Initiatives. *ICAR NEWS, A Science and Technology NewsLetter* 19(3): 4.

Saran, P. L., P. R. Kumar, R. Choudhary, G. Devi and I. S. Solanki. 2013c. New physiological disorders in papaya (*Carica papaya* L) fruit or seed production, market acceptability and economic losses in North-East Plain Zone. *Agricultural Economics Research Review* 26: 207.

Saran, P. L., I. S. Solanki and R. Rishi. 2013d. Economic impact of sole and biennial turmeric cultivation with wheat, mango and litchi as an intercrop in NEPZ. *National Seminar on Production, productivity and quality of spices*, Feburary 2–3 NRC on seed spices at Jaipur.

Satrija, F., P. Nansen, H. Bjorn, S. Murtini and S. He. 1994. Effect of papaya latex against Ascarissuum in naturally infected pigs. *Journal of Helminthology* 68(4): 343–46.

Saxholt, E., A. T. Christensen, A. Moller, H. B. Hartkopp, K. H. Ygil and O. H. Hels. 2008. Danish food composition databank. Revision 7, Department of Nutrition, National Food Institute, Technical University of Denmark. Available at http://www.foodcomp.dk/.

Schlimme, D. V. and M. L. Rooney. 1994. Packaging of minimally processed fruits and vegetables. In *Minimally Processed Refrigerated Fruits and Vegetables*, ed. R. C. Willey. New York: Chapman and Hall, 135–179.

Scotti, I. and L. F. Delph. 2006. Selective trade-offs and sex-chromosome evolution in *Silene latifolia. Evolution* 60: 1793–800.

Sepiah, M. 1993. Efficacy of propiconazole against fungi causing postharvest diseases on Eksotika papaya. *Proceedings for International Postharvest Conference on Handling Tropical Fruits*, Chiangmai, Thailand.

Seshadri, K., K. M. Usman, T. K. Kandaswamy, K. Seetharaman. 1977. Bacterial wilt of papaya caused by *Pseudomonas solanacearum. Madras Agricultural Journal* 64(3): 181–82.

Sah, H. A. and K. G. Shanmugavelu. 1975. Studies on the first generation hybrid in papaya *(Carica papaya* L.) I. morphological, foral and fruit characters. *South Indian Journal of Horticulture* 23: 100–08.

Sharma, H. C. and P. N. Bajpai. 1969. Studies on floral biology of papaya. *Indian Journal of Science and Industry* 3: 9–18.

Sharon, D., J. Hillel, A. Vainstein and U. Lavi. 1992. Application of DNA fingerprints for identification and genetic analysis of *Carica papaya* and other *Carica* spp. *Euphytica* 62(2): 119–26.

Shivannavar, A. C. 2005. An economic analysis of production and marketing of papaya in North Karnataka. Thesis, Department of Agricultural Economics, University of Agricultural Sciences, Dharwad (India).

Singfield, P. 1998. Papaya and Belize. Belize Development Trust, 19. Available at http://www.belize1.com/BzLibrary/trust19.html

Singh, B., S. K. Dwivedi and P. Eli. 1998. Studies on suitability of various structures for winter vegetable production at subzero temperatures. *Paper presented in International Horticulture Conference*, Belgium.

Singh, B., S. K. Dwivedi and J. P. Sharma. 2000. Greenhouse technology and winter vegetable production in cold arid zone. In *Dynamics of Cold and Agriculture*, eds. J. P. Sharma and A. A. Mir. New Delhi: Kalyani Publishers, 279–293.

Singh, D. B., R. K. Roshan, N. Pebam and M. Yadav. 2010. Effect of different spacings on growth, yield and yield characters of papaya (*Carica papaya* L.) cv. Coorg Honey Dew. *Acta Horticulturae* 851(1): 291–94.

Singh, I. D. 1990. *Papaya*. New Delhi: Oxford and IBH Publishing.

Singh, I. P. and C. K. Sharma. 1996. HPSC-3: A high yielding new papaya hybrid for Tripura. *Journal of Hill Research* 9(I): 73–75.

Singh, K., M. Ram and A. Kumar. 2010. Forty years of papaya research at pusa, Bihar, India. *Acta Horticulturae* 851: 81–88.

Singh, R. N. 1955. Further studies in colchicine induced polyploidy in papaya *(Carica papaya* L.). *Indian Journal of Horticulture* 12: 63–71.

Singh, R. N. 1964. Papaya breeding – A review. *Indian Journal of Horticulture* 21(2): 148–54.

Singh, S. N. 1961. Longivity of papaya pollen. *Indian Journal of Horticulture* 17(3–4): 170–75.

Sippel, A. D. and L. C. Holtzhausen. 1992. Microsporogenesis in the hermaphrodite 'Sunrise Solo' papaya (*Carica papaya* L.). *Journal of Southern African Society for Horticulture Sciences* 2: 89–91.

Sivakumar, D., N. K. Hewarathgamagae, W. R. S. Wijeratnam and R. L. C. Wijesundera. 2002. Effect of ammonium carbonate and sodium bicarbonate on anthracnose of papaya. *Phytoparasitica* 30(5): 486–92.

Smith, J. H. E. 1970. Improvement of papaya by selection and controlled pollination. *Fmg of South Africa* 46: 19–20.

Solanki, I. S., P. L. Saran and R. Rishi. 2013. Yield, soil health and economics of papaya turmeric-ginger-turnip based agri-horticultural system in Bihar. *National Seminar on Production, Productivity and Quality of Spices*, Feburary 2–3, NRC on seed spices at Jaipur, Rajasthan, India.

Somsri, S., R. J. Fletcher, R. Drew, M. Jobin, W. Lawson and M. W. Graham. 1998. Developing molecular markers for sex prediction in papaya (*Carica papaya* L.). *Acta Horticulturae* 461: 141–48.

Sondur, S. N., R. M. Manshardt and J. I. Stiles. 1996. A genetic linkage map of papaya based on randomly amplified polymorphic DNA markers. *Theoretical and Applied Genetics* 93: 547–53.

Sone, T., N. Sakamoto, K. Suga et al. 1998. Comparison of diets among elderly female residents in two suburban districts in Chiang Mai Provence, Thailand, in dry season

– survey on high- and low-risk districts of lung cancer incidence. *Applied Human Science* 17: 49–56.

Soudur, S. N., R. M. Manshardt and J. I. Stiles. 1996. A genetic linkage map of papaya on randomly amplified polymorphic DNA markers. *Theoretical and Applied Genetics* 93(4): 547–53.

Sritakae, A., P. Praseartkul, W. Cheunban et al. 2011. Mapping airborne pollen of papaya (*Carica papaya* L.) and its distribution related to land use using GIS and remote sensing. *Aerobiologia* 27(4): 291–300.

Stanghellini, H. 1993. Evapotranspiration in greenhouses with special references to Mediterranean condition. *Acta Horticulturae* 335: 295–304.

Stevenson, E. D. and C. A. Storer. 1991. Papain in organic solvents: Determination of conditions suitable for biocatalysis and the effect on substrate specificity and inhibition. *Biotechnology and Bioengineering* 37: 519–27.

Stice, K. N., A. M. McGregor, S. N. Kumar and J. Konam. 2010. Fiji red papaya: Progress and prospects in developing a major agriculture diversification industry. *Acta Horticulturae* 851: 423–26.

Stiles, J. I., C. Lemme, S. Sondur, M. B. Morshidi and R. Manshardt. 1993. Using randomly amplified polymorphic DNA for evaluating genetic relationships among papaya cultivars. *Theoretical and Applied Genetics* 85(6–7): 697–701.

Storey, W. B. 1953. Genetics of the papaya. *Journal of Hereditary* 44: 70–78.

Storey, W. B. 1958. Modifications of sex expression in papaya. *Horticulture in Advance* 2: 49–60.

Storey, W. B. 1967. Theory of derivations of the unisexual flowers of caricaceae. *Agronomia Tropical* 17: 273–321.

Storey, W. B. 1969. Papaya *(Carica papaya)*. In *Outlines of Perennial Crop Breeding in the Tropics*, eds. F. P. Ferwerda and F. Wit. The Netherlands: Wageningen, 389–407.

Storey, W. B. 1976. Papaya. In *Evolution of Crop Plants*, ed. N. W. Simmonds. London: Longman, 21–24.

Storey, W. B. 1986. *Carica papaya*. In *CRC Handbook of Flowering*, ed. A. H. Halvey, Vol. 2, Boca Raton, Florida: CRC Press Inc.

Subenthiran, S., T. C. Choon, K. C. Cheong et al. 2013. *Carica papaya* leaves juice significantly accelerates the rate of increase in platelet count among patients with dengue fever and dengue hemorrhagic fever. *Evidences Based on Complementary and Alternative Medicines* 2013: 616–737.

Subramanyam, M. D. and C. P. A. Iyer. 1981. Crossability and interspecific relationship in the genus *Carica*. *National Symposium on Tropical and Subtropical Fruit Crops*, Bangalore (India) p. 16.

Subramanyam, M. D. and C. P. A. Iyer. 1986. Flowering behaviour and floral biology in different species of *Carica* genus. *Haryana Journal of Horticulture Sciences* 15(3–4): 179–87.

Sudha, R., D. R. Singh, M. Sankaran, S. Singh, V. Damodaran and P. Simachalam. 2012. Genetic diversity analysis of papaya (*Carica papaya* L.) genotypes in Andaman Islands using morphological and molecular markers. *African Journal of Agriculture Research* 8(41): 5187–91.

Sukhada, M. 1992. Effect of VAM inoculation on plant growth, nutrient level and root phosphatase activity in papaya (*Carica papaya* C.V. Coorg Honey Dew). *Fertilizer Research* 31: 263–67.

Tanwar, R. K., P. Jeyakumar and S. Vennila. 2010. Papaya mealybug and its management strategies. Technical Bulletin 22, *National Centre for Integrated Pest Management*, New Delhi.

Tapia, C. E., E. M. A. Gutierrez, L. M. Warbourton, A. D. Uriza and M. A. Rebellodo. 2005. Characterization of pineapple germplasm (*Ananas* spp) by mean AFLPs. *Proceedings in Fourth International Pineapple Symposium. Acta Horticulturae* 666: 109–14.

Teixeira da Silva, J. A., Z. Rashid, D. T. Nhut et al. 2007. Papaya (*Carica papaya* L.) biology and biotechnology. *Tree Forest Science and Biotechnology* 1(1): 47–73.

Tennant, P. F., M. H. Ahmad and D. Gonsalves. 2002. Transformation of *Carica papaya* L. with virus coat protein genes for studies on resistance to papaya ringspot virus from Jamaica. *Tropical Agriculture (Trinidad)* 79(2): 105–13.

Thompson, A. K. 2001. *Controlled Atmosphere Storage of Fruits and Vegetables*. UK: CAB International.

Thumdee, S., A. Manenoi and R. E. Paull. 2007. Activity of papaya fruit hydrolases during natural softening and modified softening. *Acta Horticulturae* 740: 317–22.

Tigist, M., V. M. Rao and G. Faye. 2014. Determination of essential and non-essential metals concentration in papaya (*Carica papaya*) seeds, leaves and supporting soil of odoshakiso district in South East Oromia region, Ethiopia. *International journal of Research in Pharmacy and Chemistry* 4(1): 202–16.

Tripathi, S., J. N. Y. Suzuki, S. A. Ferreira and D. Gonsalves. 2008. Papaya ringspot virus-P: Characteristics, pathogenicity, sequence variability and control. *Molecular Plant Pathology* 9(3): 269–80.

Trivedi, R. K. and M. Gunasekaran. 2013. Indian minimum seed certification standards. *The Central Seed Certification Board, Department of Agriculture & Co-operation Ministry of Agriculture,* Government of India, New Delhi, p. 605.

Tsai, S. F., S. D. Yeh, C. F. Chan and S. I. Liaw. 2009. High-efficiency vitrification protocols for cryopreservation of *in vitro* grown shoot tips of transgenic papaya lines. *Plant Cell, Tissue and Organ Culture* 98: 157–64.

Tsay, J. G., R. S. Chen, H. L. Wang, W. L. Wang and B. C. Weng. 2011. First report of powdery mildew caused by *Erysiphe diffusa*, *Oidium neolycopersici*, and *Podosphaera xanthii* on papaya in Taiwan. *Plant Diseases* 95: 1188.

Tsuge, H., T. Nishimura, Y. Tada et al. 1999. Inhibition mechanism of cathepsin L specific inhibitors based on the crystal structure of papain-CLIK148 complex. *Biochemistry and Biophysics Research Communication* 266: 411–16.

Uche-Nwachi, E. O., C. V. Mitchell and C. McEwen. 2011. Steroidogenic enzyme histochemistry in the testis of Sprague Dawley rats following the administration the water extracts from *Carica papaya* seed. *African Journal of Traditional Complementry and Alternative Medicines* 8(1): 69–78.

Udoh, P., I. Essien and F. Udoh. 2005. Effects of *Carica papaya* (paw paw) seeds extract on the morphology of pituitary-gonadal axis of male Wistar rats. *Phototherapy Research* 19(12): 1065–68.

Ulloa, J. B., J. H. van Weerd, E. A. Huisman and J. A. J. Verreth. 2004. Tropical agricultural residues and their potential uses in fish feeds: The costarican situation. *Waste Management* 24: 87–97.

Urasaki, N., M. Tokumoto and K. Tarora et al. 2002. A male and hermaphrodite specific RAPD marker for papaya (*Carica papaya* L.) *Theoretical and Applied Genetics* 104: 281–85.

USDA (United States Department of Agriculture), 2009. Agricultural Research Service, National Nutrient Database for Standard Reference, Release 22, Nutrient Data Laboratory Home Page. Available at http://www.ars.usda.gov/ba/bhnrc/ndl.

Van Kampen, V., R. Merget and T. Bruning. 2005. Occupational allergies to papain. *Pneumologie* 59: 405–10.

Vega-Frutis, R. and R. Guevara. 2009. Different arbuscular mycorrhizal interactions in male and female plants of wild *Carica papaya* L. *Plant Soil* 322(1): 165–76.

Verma, A. K. 1996. Viral and mycoplasmal disease of papaya *(Carica papaya* L.). In *Disease Scenario in Crop Plants*, eds. V. P. Agnihotri, O. Prakash, K. Ram and A. K. Mishra. Delhi: International Books and Periodicals Supply Co, 115–135.

Vidhyasagar, G. M. and S. M. Murthy. 2013. Medicinal plants used in the treatment of Diabetes mellitus in Bellary district, Karnataka. *Ind J Trad Knowledge* 12(4): 747–51.

Vig, A. P., G. Rampal, T. S. Thind and S. Arora. 2009. Bio-protective effects of gluosinolates-A review. *Food Science and Technology* 42: 1561–72.

Vivas, M., S. F. D. Silveira, C. E. P. D. S. Terra and M. G. Pereira. 2011. Testers for combining ability and selection of papaya hybrids resistant to fungal diseases. *Crop Breeding and Applied Biotechnology* 11: 36–42.

Volden, J., G. I. A. Borge, G. B. Bengtsson, M. Hansen, I. E. Thygesen and T. Wicklund. 2008. Effect of thermal treatment on glucosinolates and antioxidant-related parameters in red cabbage *(Brassica oleracea* L. ssp. *capitataf rubra). Food Chemistry* 109: 595–605.

Volden, J., G. I. A. Borge, G. B. Bengtsson, M. Hansen, T. Wicklund and G. B. Bengtsson. 2009. Processing (blanching, boiling, steaming) effects on the content of glucosinolates and antioxidant-related parameters in cauliflower (*Brassica oleracea* L. ssp. *botrytis*). *Food Science and Technology* 42: 63–73.

Wai, C. M., R. Ming, P. H. Moore, R. E. Paull and Q. Yu. 2010. Development of chromosome-specific cytogenetic markers and merging of linkage fragments in papaya. *Tropical Plant Biology* 3: 171–81.

Wall, M. M. 2006. Ascorbic acid, vitamin A and mineral composition of banana (*Musa* sp.) and papaya (*Carica papaya*) cultivars grown in Hawaii. *Journal of Food Composition and Analysis* 19: 434–45.

Walter, L. 2008. Cancer remedies. Available at Health-science-sprite.com/cancer6-remedies.

Wang, D. and W. H. Ko. 1975. Relationship between deformed fruit disease of papaya and boron deficiency. *Phytopathology* 65(4): 445–47.

Wang, J., C. Chen, J. K. Na et al. 2008. Genome-wide comparative analyses of microsatellites in papaya. *Tropical Plant Biology* 1: 278–92.

Wang, J., J. K. Na, Q. Yu et al. 2012. Sequencing X and Yh chromosomes in papaya revealed the molecular basis of incipient sex chromosome evolution. *Proceedings of National Academy of Science of United State of America* 109(34): 13710–15.

Webman, E. J., G. Edlin and H. F. Mower. 1989. Free radical scavenging activity of papaya juice. *International Journal of Radiation and Biology* 55(3): 347–51.

Wicker, C. A., R. P. Sahu, K. K. Datar, S. K. Srivastava and T. L. Brown. 2010. BITC sensities pancreatic adenocarcinomas to TRAIL-induced apoptosis. *Cancer Growth Metastasis* 2: 45–55.

Wills, R. B. H., J. S. K. Lim and H. Greenfield. 1986. Composition of Australian foods -31. Tropical and Sub-tropical Fruit. *Food Technology of Australia* 38: 118–23.

Wills, R. B. H., W. B. McGlasson, D. Graham, T. H. Lee and E. G. Hall. 1989. *Postharvest-An Introduction to the Physiology and Handling of Fruit and Vegtables*, 3rd edn., New York, U.S.A.: Van Nostrand Reinhold, p. 46.

Wood, C. B., H. W. Pritchard and D. Amritphale. 2000. Desiccation-induced dormancy in papaya (*Carica papaya* L.) seeds is alleviated by heat shock. *Seed Science Research* 10: 135–45.

Workneh, T. S., M. Azene and S. Z. Tesfay. 2012. A review on the integrated agro-technology of papaya fruit. *African Journal of Biotechnology* 11(85): 15098–110.

Wright, J. W. and T. R. Meagher. 2003. Pollination and seed predation drive flowering phenology in *Silene latifolia* (Caryophyllaceae). *Ecology* 84: 2062–73.

Xi, Z. Y., F. H. He, R. Z. Zeng et al. 2008. Characterization of donor genome contents of backcross progenies detected by SSR markers in rice. *Euphytica* 3: 369–77.

Xian, M., X. Chen, Z. Liu, K. Wang and P. G. Wang. 2000. Inhibition of papain by *S*-nitrosothiols. *Journal of Biological Chemistry* 275: 20467–73.

Xiao, D., A. A. Powolny and S. V. Singh. 2008. Benzyl isothiocynate targets mitochondrial respiratory chain to trigger reactive oxygen species-dependent apoptosis in human breast cancer cells. *The Journal of Biological Chemistry* 283: 30151–63.

Yadav, A. L. and J. Prasad. 1990. Studies on sex-inheritance of papaya hybrids *(Carica papaya* L.). *Indian Journal of Horticulture* 47(4): 385–88.

Yahner, R. H. and A. L. Wright. 1985. Depredation on artificial ground nests: Effects of edge and plot age. *Journal of Wildlife Management* 49: 508–13.

Yang, J., T. Yu, Y. Cheng and S. Yeh. 1996. Transgenic papaya plants from *Agrobacterium* mediated transformation of petioles of *in vitro* propagated multishoots. *Plant Cell Reports* 15(7): 459–64.

Yang, J. S. and C. A. Ye. 1992. Plant regeneration from petioles of *in vitro* regenerated papaya (*C. papaya* L.) shoots. *Botanical Bulletin-Academia Sinica* 33(4): 375–81.

Yeh, S., H. Bau, Y. Cheng, T. Yu and J. Yang. 1998. Greenhouse and field evaluation of coat-protein transgenic papaya resistant to papaya ringspot virus. *Acta Horticulturae* 461: 321–28.

Yogananda, D. K., B. S. Vyakaranahal and M. Sekhargouda. 2004. Effect of seed invigoration with growth regulators and micronutrients on germination and seedling vigour of bell pepper cv. California wonder. *Karnataka Journal of Agricultural Sciences* 17(4): 811–13.

Yogeesha, H. S., K. Bhanuprakash and L. B. Naik. 2008. Seed storability in three varieties of papaya in relation to seed moisture, packaging material and storage temperature. *Seed Science and Technology* 36(3): 721–29.

Yousefi, A., M. Niakousari and M. Moradi. 2013. Microwave assisted hot air drying of papaya (*Carica papaya* L.) pretreated in osmotic solution. *African Journal of Agriculture Research* 8(25): 3229–35.

Yu, Q., S. Hou, R. Hobza et al. 2007. Chromosomal location and gene paucity of the male specific region on papaya Y chromosome. *Molecular Genetics and Genomics* 278: 177–85.

Yu, Q., S. Hou, F. A. Feltus et al. 2008a. Low X/Y divergence in four pairs of papaya sex-linked genes. *The Plant Journal* 53: 124–32.

Yu, Q., R. Navajas-Perez, E. Tong, J. Robertson and P. H. Moore. 2008b. Recent origin of dioecious and gynodioecious Y chromosomes in papaya. *Tropical Plant Biology* 1: 49–57.

Yu, Q., E. Tong, R. L. Skelton et al. 2009. A physical map of the papaya genome with integrated genetic map and genome sequence. *BMC Genomics* 10: 371.

Yu, T. A., S. D. Yeh and J. S. Yang. 2003. Comparison of effects of kanamycin and geneticin on regeneration of papaya from root tissue. *Plant Cell, Tissue and Organ Culture* 74: 169–78.

Yung, J. S. 1986. Interspecific hybridization and immature embryo culture of *Carica* sp. *Journal for College of Science and Engineering, National Chung Hsiang University, Taichung, Taiwan* 23: 13–26.

Zerpa, D. M. 1957. Triploides De *Carica papaya* L. *Agronomica Tropical* 7: 83–86.

Zerpa, D. M. 1959. Cytology of inter-specific hybrids in *Carica. Agronomica Tropical Venezuela* 8: 135–44.

Zhang, L. X. and R. E. Paull. 1990. Ripening behavior of papaya genotype. *Horticulture Science* 25: 454–55.

Zhang, W., X. Wang, Q. Yu, R. Ming and J. Jiang. 2008. DNA methylation and heterochromatinization in the male-specific region of the primitive Y chromosome of papaya. *Genome Research* 18: 1938–43.

Zhang, W., C. M. Wai, R. Ming, Q. Yu and J. Jiang. 2010. Integration of genetic and cytological maps and development of a pachytene chromosome-based karyotype in papaya. *Tropical Plant Biology* 3: 166–70.

Zhou, L., D. A. Christopher and R. E. Paull. 2000. Defoliation and fruit removal effects on papaya fruit production, sugar accumulation, and sucrose metabolism. *Journal of American Society for Horticulture Sciences* 125(5): 644–52.

Zhou, K., H. Wang, W. Mei, X. Li, Y. Luo and H. Dai. 2011. Antioxidant activity of papaya seed extracts. *Molecules* 16(8): 6179–92.

Zluvova, J., S. Georgiev, B. Janousek, D. Charlesworth, B. Vyskot and I. Negrutiu. 2007. Early events in the evolution of the *Silene latifolia* Y chromosome: Male specialization and recombination arrest. *Genetics* 177: 375–86.

Zulhisyam, A. K., T. S. Chuah, A. I. Ahmad, N. N. Azwanida, S. Shazani and M. H. Jamaludin. 2013. Effect of storage temperature and seed moisture contents on papaya (*Carica papaya* L.) Seed viability and germination. *Journal of Sustainability Science and Management* 8(1): 87–92.

Index

A

B

C

D

J

K

L

M

N

O

P

R

S

T

U